Praise for *Spineless*

"*Spineless* is as mesmerizing, surprising, and beautiful as the jellyfish itself. Every page contains some astonishing treasure. If you cherish the sea, if you care about the environment, if you relish life on this sweet, blue planet, you will love this book."

—Sy Montgomery, *New York Times* bestselling author of *The Soul of an Octopus*

"Part travelogue, part memoir, part deep-dive (literally) into the world of jellyfish. . . . *Spineless* can serve as inspiration for any of us to reclaim a creative space in the midst of family life."

—NPR

"[*Spineless* is] fascinating. . . . This combination of insider and outsider perspective is uniquely suited to a book on creatures whose internal organs are visible through transparent outer layers."

—*Wall Street Journal*

"[Berwald's] sense of wonder is infectious and the book is a heartfelt plea for humans to fulfill their responsibilities toward nature."

—*The New Yorker*

"In this memoir/science-reporting mash-up, [Berwald] profiles one of the ocean's most intriguing creatures—the unique contractions it uses to propel through water, its acidifying habitat, and its booming populations."

—*Scientific American*

"An astonishingly gorgeous book. . . . Science enthusiasts, curious animal lovers, and those who want to educate themselves more on climate change's effect on our oceans will find this book irresistible."

—*W Magazine*

"Berwald's clear, delectable, and accessible prose . . . forces the reader to reconsider the future of our planet, and of our role in it. It will, at the very least, leave you with a newfound appreciation of the translucent, spineless jellies."

—*BuzzFeed*

"A wandering, compelling mixture of memoir and nature writing . . . [written with] clear, strong prose, a welcome help when dealing with the complex stew of biomechanics, chemistry, and evolutionary theory that the subject entails."

—*The Texas Observer*

"There is perhaps no more soothing sight than the illuminated jellyfish tanks in an aquarium. In *Spineless* Juli Berwald brings us inside, unraveling a memoir about the scientific exploration of these strange, wonderful creatures."

—*Popular Science*

"Berwald offers an engrossing look at the enigmatic sea creature most easily recognised in its swimming 'medusa' form. . . . A revelatory science memoir."

—BBC Culture

"Berwald doesn't rebut the dark jellyfish narrative, but she usefully qualifies it, exploring a diversity of jellyfish responses to harms unevenly distributed throughout the sea. There is no global jellyfish ecophagy. The real bloom, Berwald argues, is in jellyfish science, where the interplay of jellyfish and their ecosystems is only now beginning to be pieced together." —*The Atlantic*

"It's a story of personal discovery, rediscovery of the underwater world, and an earth-spanning journey to study these complex creatures, all the while throwing into stark relief the importance of understanding and protecting our increasingly endangered marine ecosystems." —*Southern Living*

"Fascinating. . . . Readers can't help but be swept away with enthusiasm. . . . Full of humor and intrigue, *Spineless* is a seaworthy saga brimming with information about not only jellyfish but also about the health and future of the oceans and our planet." —*BookPage*

"*Spineless* explores not only jellies' suffering ecosystem and our responsibility to the planet but also how Berwald strengthened her own backbone." —*Real Simple*

"Stunning memoir." —*Bustle*

"Captivating and informative." —*Publishers Weekly* (starred review)

"In this lovely exploration of the mysterious jellyfish, Berwald both entrances and sounds a warning: pay attention to the messages sent by ocean life, and act to protect their environment, and ours." —*Kirkus Reviews*

"In this astonishing adventure of a book, Juli Berwald takes us on a personal journey into the enchanting and mystifying aqueous world of jellyfish, and in so doing, sheds light on the vital ecological balances upon which our own survival depends."

—Ruth Ozeki, author of *A Tale for the Time Being*

"Like Bernd Heinrich, Juli Berwald illuminates an entire ecosystem through the lens of one remarkable group of creatures. I learned something from every page, not only about jellyfish but about the crucial role of marine scientists and their studies."

—Andrea Barrett, author of *Archangel* and *The Voyage of the Narwhal*

"Jellyfish have captured the imagination of today's mass public, and Juli Berwald in *Spineless* spins a tale of almost magical science and eye-opening evolutionary history in telling us why we are so darn fascinated by these globby creatures. They can swim on their own propulsion. They can kill with alarming efficiency. They dazzle with their

shapes and hues, taking their place as the eccentric beauties of our oceans. Berwald makes science a page-turner. I read the book in one sitting, as if *Spineless* were the hit summer mystery."
—Diana Nyad, author of ***Find a Way***

"With lucid, gorgeous, arresting language, Juli Berwald opens a rare window on the adventures of scientists at work, capturing the fabulously ingenious ways they make sense of our world. *Spineless* will leave you awestruck, revering the ever-more fragile brilliance of life on Earth."
—Miriam Horn, author of ***Rancher, Farmer, Fisherman***

"Unexpected and uniquely delightful."
—Parade

"Partially an environmental investigation and partially a memoir, *Spineless* is a book that will make you feel smarter and not only more aware of your surroundings, but more aware of yourself."
—HelloGiggles

"She swam with jellies, watched how quickly they disintegrate in fishers' nets, ate them in Japan, and kept them in a home aquarium, and as she revels in these spineless animals, she teaches us to delight in them, too."
—Booklist

"It morphs with disarming ease from memoir complete with wacky characters, sharp dialogue, and unexpectedly touching moments to natural history infused with curiosity and no small degree of wonder. . . . The 'jellyfish journey' of *Spineless* is strange and ultimately quite charming."
—Open Letters Monthly

"Berwald excels at depicting the wonder and appreciation she has gained for the strange, gelatinous creatures and the ocean that sustains them. . . . [Jellyfish] are fascinating in part because there's so much more to find out about them. In *Spineless*, Berwald demonstrates that our oceans represent a scientific frontier at least as exciting and promising as space, and posits jellyfish as a prime candidate for study and appreciation."
—Shelf Awareness

"More than a mere look at jellyfish. . . . [*Spineless*] provides plenty of food for thought for those seeking to make different choices in their treatment of the environment and in their own lives."
—Library Journal

Spineless

THE SCIENCE OF JELLYFISH AND

THE ART OF GROWING A BACKBONE

Juli Berwald

Illustrations by Rachel Ivanyi

RIVERHEAD BOOKS

New York

RIVERHEAD BOOKS
An imprint of Penguin Random House LLC
375 Hudson Street
New York, New York 10014

The Library of Congress has catalogued the Riverhead hardcover edition as follows:

Names: Berwald, Juli, author.
Title: Spineless : the science of jellyfish and the art of growing a backbone / Juli Berwald ; illustrations by Rachel Ivanyi.
Description: New York : Riverhead Books, 2017. | Includes bibliographical references and index.
Identifiers: LCCN 2017005838 (print) | LCCN 2017007895 (ebook) |
ISBN 9780735211261 (hardcover) | ISBN 9780735211278 (ebook)
Subjects: LCSH: Jellyfishes.
Classification: LCC QL377.S4 B47 2017 (print) | LCC QL377.S4 (ebook) | DDC 593.5/3—dc23
LC record available at https://lccn.loc.gov/2017005838
p. cm.

First Riverhead hardcover edition: November 2017
First Riverhead trade paperback edition: November 2018
Riverhead trade paperback ISBN: 9780735211285

Printed in the United States of America

Book design by Gretchen Achilles
Illustrations by Rachel Ivanyi

Penguin is committed to publishing works of quality and integrity. In that spirit, we are proud to offer this book to our readers; however, the story, the experiences, and the words are the author's alone.

For Keith

Contents

Planula

1. If You Dare 3

Polyp

2. What's Your Agenda? 21
3. Jellyfish Salad 41
4. Missing Polyp 53
5. In Jelly Genes 67
6. Robojelly 83
7. Seeing What's Not There 95
8. Day-glo Jellies 109
9. Jellyfish Sense 121
10. The Nerve of the Jellyfish 131
11. Life's Limits 147

Strobila

12. The Bottom of the Wave 165

Ephyra

13. Stop Waiting 183
14. Sacred Island 197
15. Stalking the Beast 207
16. Jellyfish al Dente 215
17. Jellyfishing 225

Medusa

18. Toxic Cocktail 243
19. Sting Block 257
20. In Medusa's Blood 267
21. Party Like a Jellyfish 283
22. Bloom 295

Acknowledgments 307
Notes 311
Index 327

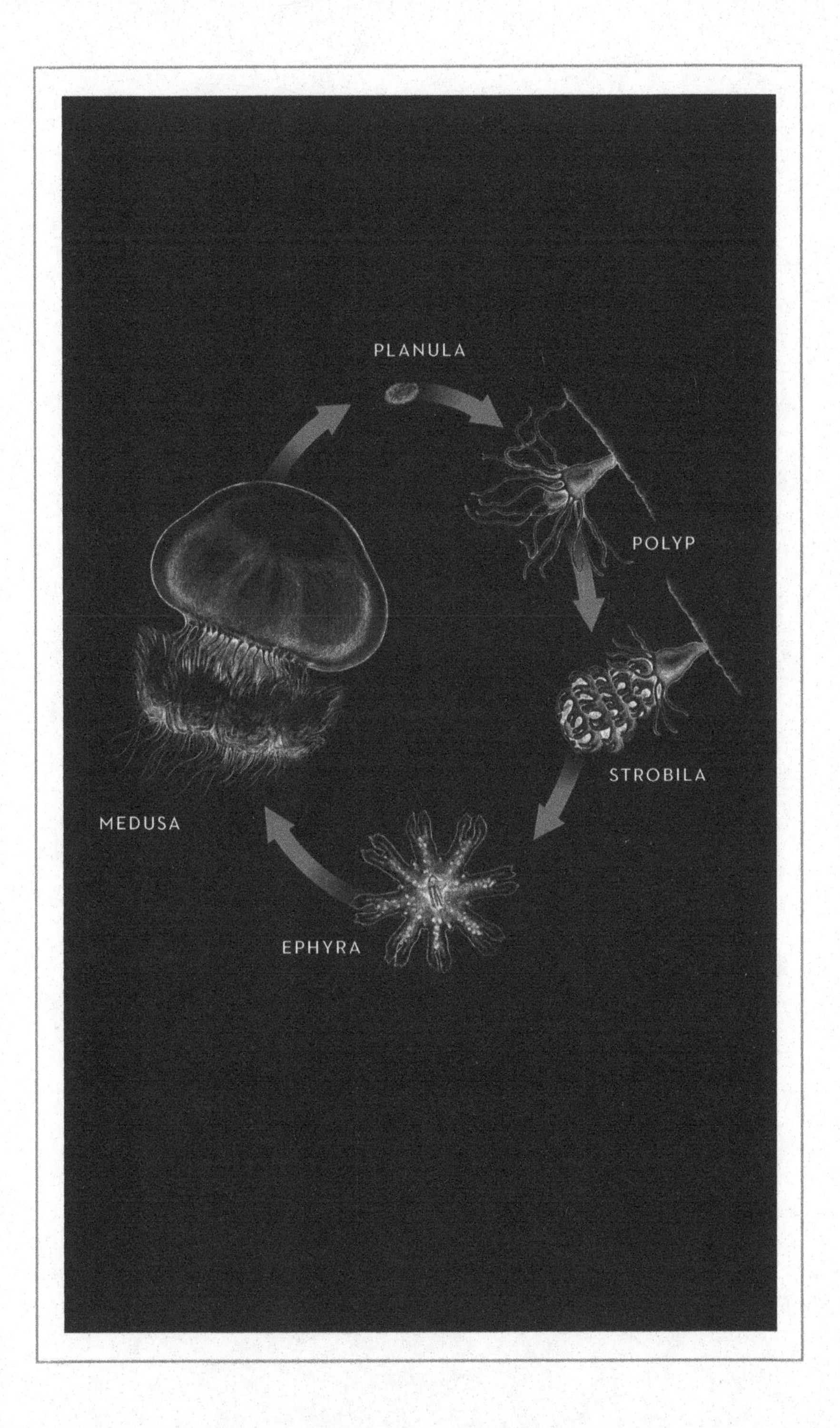
PLANULA
POLYP
STROBILA
EPHYRA
MEDUSA

Among the most beautiful, as well as the most common, of the marine animals which are to be met with upon our coasts are the jelly-fish. . . . And in the memories of most of us is there not associated with the picture of breaking waves and sea-birds floating indifferently in the blue sky or on the water still more blue, the thoughts of many a ramble among the weedy rocks and living pools, where for the time being we all become naturalists, and where those who least know what they are likely to find in their search are most likely to approach the keen happiness of childhood? If so, the image of . . . those crystal globes pulsating with life and gleaming with all the colors of the rainbow . . . are perhaps the most strange, and certainly in my estimation the most delicately lovely creatures in the world.

GEORGE J. ROMANES, *Jelly-fish, Star-fish and Sea-Urchins: Being a Research on Primitive Nervous Systems,* 1885

Planula

If You Dare

Hiroshima's downtown is a garden of modern architecture interspersed with swaths of lovely green parks. In the center, there is a single structure, in ruins, capped by a skeleton of curved iron. This is the Atomic Bomb Dome, located at the destruction's epicenter, the sole building that managed to remain standing amid the massive force that flattened everything else for miles in all directions. It is an astonishing memorial to both our capacity for horrifying devastation and our awesome resilience. The dome sits along the side of one of six tidal streams that flow through Hiroshima. In the murky, green water, I watched thousands—maybe hundreds of thousands—of pale pink disks parade by, a flood of jellyfish. Juxtaposed with the dome, the endless stream of jellyfish seemed to square off nature's power against our own, a battle as old as civilization that continues to play out in the decisions we make today. They were the first wild jellyfish I saw in years of chasing jellyfish. The milky creatures pulsed slowly, slower than my heartbeat, which dropped as I watched. The movements of their bells trailing gossamer tentacles were like millions of eyelashes blinking open and closed and open again, giving me a feeling that these alien animals could peer deep into the soul of the sea. I found it impossible to fathom the source of this endless river of life. The jellies continued to flow by for as long as I stood and watched.

Maybe you remember the first time you saw a jellyfish. Maybe you were lucky enough to stick your head under the ocean's surface, wearing a mask, and watch the primal undulations of a jellyfish dancing to some internal rhythm. Maybe you felt a biting sting in the water and turned to see a gelatinous blob disappear like sinking mucus. Maybe you stood against the glass and watched the elegant clover on the back of a moon jellyfish or the graceful train of a sea nettle in an aquarium. Or maybe it was footage of jellyfish hawking a cure for memory loss.

I don't remember the very first jellyfish I saw. I grew up in a time before aquaria were full of jellyfish tanks, in a very landlocked in St. Louis. I never spent an extended period near the ocean until I was in college. In my junior year, 1987, I attended an English-speaking study-abroad program in Tel Aviv. From a New England school where getting ready to go to a party meant putting on tattered jeans, a flannel shirt, and duck boots, I was a misfit among the other American students, who were a lot less style-challenged. I found the Israelis my age no easier to get along with. At nineteen, they were required to serve in the army, while I had the freedom to jaunt off to college unencumbered.

I stuffed my loneliness down with the hummus and baba ghanoush sold from a cart on the street corner near my cinder-block dorm. At about the same rate that my weight increased, my spirits sank. One day, I opened the door to the building where my class on Middle Eastern politics was held and spotted a sign out of the corner of my eye. It advertised a marine biology course: a week in Eilat, at the southern tip of Israel, studying the ecology of the Red Sea. Knowing next to nothing about biology or Eilat, not to mention the fact that I was hardly able to squeeze into my swimsuit anymore, I scrawled my name at the bottom of the list. It was clear I needed to put some distance between me and the pita cart.

A few days later, with about twenty other students, I boarded a bus headed south at dawn. After a five-hour ride through the desert, we were offloaded and handed snorkels, masks, and fins. I think there was a quick instruction on keeping hands away from urchin spines and fire coral. I shuffled down the beach backward, the only way to walk in fins. When I slipped my face below the water's surface, it was as if I were Dorothy stepping into Oz. My dull, sad world erupted in a kaleidoscope of colors, shapes, and textures. I glided over creatures that I could never have imagined existed in my mostly inland life. Pink, purple, and yellow coral that branched, encrusted, and mounded. Blotchy lavender anemones, creeping crimson shrimp, twirling buff-colored worms, filigreed fans, spiked black urchins. A day-glo orange sponge with a violet slug curled around it. Bowls of lemon-, lime-, and watermelon-hued clams. Here, astonishingly, was paradise.

That evening, the professor gave a lecture detailing twenty different species of coral living on the reef. He told us to memorize them all. Walking to the dorms after class, gazing at the moon's reflection on the Red Sea, I felt, for the first time in a long time, happy.

Within days, I learned not only to recognize and identify the twenty different coral species, but to understand how they got along on the reef. I laid out thirty-foot lengths of rope and measured the space each colony occupied along the line. I watched coral engage in territorial feuds when their delicate tissues grew too close together. I learned to recognize the winner as the one whose threadlike stinging cells outnumbered those of its foe. It was a contrast to the battles that destabilized the Middle East: I found truth and beauty in these miniature clashes, which brought balance to the ecosystem.

By the time I climbed back on the bus to Tel Aviv, marine biology had sunk its hooks deep into me. I felt that the intricate dynamics of the ocean were more deeply authentic than anything humans could create. Seven days of snorkeling and studying coral all day and then peering into microscopes at constellations of plankton deep into the night taught me that biology is the foundation of our world, the stuff that is not changed with the whims of humans. Its heritage, so much more ancient than our own, gave it the gravitas I'd been searching for. Studying this scaffolding on which we hang our existence was what I wanted to do with my life.

After that first mind-blowing dive in Israel, I did go on to study ocean science in graduate school. And that's probably the first time I saw a jellyfish. A few years after my work on jellyfish started, I dug out my old box of grad-school memorabilia. And in a cardboard clamshell of old slides, the kind you have to stick into a projector, I found a hazy picture of what I now know was a compass jellyfish, taken in Puget Sound. I don't remember taking it. In grad school, I didn't work on jellyfish or any other marine animal. I worked on the equations satellites use to tell us how much carbon dioxide the ocean can inhale. That was the big picture, anyway, the kind of thing I told my parents so they could repeat it to their friends. The reality was much more obscure. I studied a single variable in those equations, one that only about a dozen people in the world recognized.

After moving to landlocked Austin, Texas, with my husband, I glumly slipped off the academic path and found my way into writing textbooks. It worked well when my two kids were young and I was always sleep deprived. But state-mandated academic standards soon quashed any sense of passion about the work. I had written and rewritten the same chapters on the physics of tossing balls into the air and dropping rocks off cliffs, on the bending and reflecting of waves, for nearly a decade when a friend gave my name to an editor at *National Geographic.* They needed a writer to pen three paragraphs on the physics of water bugs skating up a meniscus. After it was done, I spotted the magazine on a grocery store rack. Flipping to my tiny byline to show the cashier, I crowed, "That's me!"

Despite my small successes in the print media and my semi-stable textbook gigs, when I slowed down enough to think, I felt lost. Although I looked good on paper, and maybe Facebook too, I knew I was actually adrift, aimlessly riding currents over which I had no control.

The technical name for the stage of a jellyfish's life when it swims freely in the seas is medusa, a moniker shared with the ancient Greek mythological monster. Medusa is famous for her horrible face, which could turn a man to stone, and her wild locks of hissing snakes. It's not hard to see the similarities. A swimming medusa could look like a floating head with a wayward mane of terrifying stinging tentacles.

But dig a little deeper into the story of Medusa, and what you find is not at all a monster, but a victim whose story has been misunderstood. Medusa was born to two ancient marine deities and, according to Ovid, was stunningly gorgeous. She served the goddess Athena in her temple. Some say she was a temptress and lured Poseidon into Athena's temple. Others say Poseidon couldn't control himself. As in too many cases like this, it depends who's telling the story. Since I am: He raped her, right there in Athena's temple.

Because it was ancient Greece, Medusa had little recourse. Athena went into a rage that her temple had been defiled, turned Medusa into a hideous monster, and kicked her out. Medusa—who deserved at least a shoulder to cry on, if not justice in a court—instead banished herself to a remote island, frightened and deformed. Imagine her disgust, shock, and horror to find that she was also pregnant with her assailant's offspring. To complete the tragedy, the man always regarded as a hero, Perseus, arrived. Using the tricks of the gods—an invisibility cloak, winged sandals, and a mirrored shield—he snuck up on Medusa. She never had a chance. He murdered her.

From her severed neck, twin newborns emerged. The first was Pegasus, the winged horse, who flew up to the heavens. The other was Chrysaor, a giant who carried a golden sword. Today, a group of golden-brown jellyfish species known as sea nettles, some of which can grow to three feet across, is among the more common jellyfish groups in the oceans. Their name honors poor Medusa's child. The genus is called *Chrysaora*.

Like their mythical namesake, jellyfish are also awash in misunderstanding. They have no centralized brain, but they see and feel and react to their environment in complex ways. Their body form looks simple, yet their swimming ability is the most economical in the animal kingdom. We know them in their swimming medusa form, but they live as much or more of their lives as a mysterious tiny tube planted on the underside of a rock. They wash ashore in hordes, and they dominate the deepest depths of our planet, supporting entire ecosystems. And still, scientists are unable to predict where and when to find them. To some, jellyfish symbolize the monster—not only in their potentially lethal sting, but also in a more global sense, as a symptom of ecosystem demise. At the same time, jellyfish are utterly, breathtakingly beautiful. As we watch the primal undulations of a medusa, they harmonize with the beats of our own hearts. Perhaps the extraordinary creativity of this balance, this surprising ability to exist in the space between monster and goddess, is why jellyfish resonate so deeply within all of us. Perhaps the story of jellyfish is really about our own possibilities.

I stumbled onto jellyfish, not while snorkeling or visiting an aquarium but while working with *National Geographic* photographer David Liittschwager. David is recognized for his close-up shots of marine organisms nearly always posed on a simple background of white or black. Once, when I asked him about the style, he said he used the plain background because it allows the observer to regard the animal as an individual. And it does. Staring at a shot of a solitary snail whose dark eyes peer out from under a pearly white shell or a nearly transparent octopus whose arms are coiled in symmetric spirals, you realize that every living creature on our planet tells a unique millennia-long success story worthy of our consideration.

The piece David was working on at the time was on ocean acidification, which is known as global warming's evil twin. The oceans have sucked up about 28 percent of the carbon dioxide we've emitted from burning fossil fuels over the past two and a half centuries. When carbon dioxide mixes with water, it forms carbonic acid. With more carbon dioxide in the atmosphere, there's more carbonic acid in the ocean. Acid dissolves shells, and many marine creatures have shells: coral, crabs, snails, lobster, shrimp, urchins, sea stars, even some worms. The ocean is now 30 percent more acidic than it was when the first Model T rolled off the assembly line a century ago.

As I scanned the story layout, I saw a quintessential *National Geographic* chart called "Winners and Losers." On the losers' side were creatures with shells: lobsters, crabs, sea stars. That made sense. Shells dissolve in acid seas.

On the winners' side were shell-less things: algae, worms, and jellyfish. Jellyfish? I'd always thought of jellyfish as delicate. What was their protection against acidified surroundings? I was curious. Delving into scientific literature, I found that drawing connections between jellyfish and today's more acidic seas was not simple. Strikingly few experiments on jellyfish in future ocean conditions had been done. A few studies had shown that warmer temperatures increased the growth of the planted tube part of the jellyfish life cycle called the polyp. One study had shown that these polyps fared poorly in water with low pH. One had shown they could survive polluted conditions. But that seemed meager evidence. I delved deeper.

I found fascinating stories about jellyfish all over the globe. Beginning in 1963, off Namibia in southwestern Africa, jellyfish started dominating the ecosystem, depressing what had been one of the world's most productive fisheries. Today, those waters are a sea of goo. In the Black Sea, populations of a comb jelly imported from the Atlantic in ballast water exploded into the millions, causing hundreds of millions of dollars in economic losses until its predator, another comb jelly from the Atlantic, was imported too. In Japan, the presence of the giant jellyfish was recorded only three times during the twentieth century. It has appeared regularly in the twenty-first, and in 2009 accounted for a massive aggregation of jellyfish, known as a bloom, the largest in recorded history. In the Philippines, jellyfish clogged the cooling system of a major power plant, causing a blackout that was mistaken by some as the start of a military coup. The largest lagoon on Spain's Mediterranean coast, called the Mar Menor, had always been home to jellyfish, but in the 1990s the numbers inflated. The jellyfish grew so thick in 2002 that you couldn't drive a boat through the water.

Were these changes a symptom of a changing planet? Or were they just natural cycles that we'd failed to see because we weren't looking? Answering that question is complicated because jellyfish have been historically understudied by scientists. Except for rare allocations, such as the Jellyfish Act of 1966, which provided funding to study sea nettles in the Chesapeake, there had not been concerted effort to understand jellyfish and their roles in ecosystems.

Tempting as it was to draw conclusions, some scientists cautioned that we might not have enough data to do so. In a 2008 paper on jellyfish and acidification, Steve Haddock, a jellyfish researcher at the Monterey Bay Research Institute in California, chastened other researchers who were drawing connections between jellyfish and the climate. "Broad statements about global warming and climate change are likely to garner a great deal of attention, and thus should be held to a relatively high standard of analytical rigor . . ." he wrote. "Because research of this kind can have lasting policy implications, it is important not to attribute all perceived ecological change to 'global warming' or 'ocean acidification' without reasonable evidence from an appropriately broad data set."

Amid the usual staid scientific results and discussion, here was language that appealed to scientists to steer clear of sensationalizing, even as the words acknowledged that jellyfish were gaining some sort of new scientific status. The passage made me think jellyfish had an important story to tell. As it turns out, they have many.

When I first learned about jellyfish as a student in the 1980s, they were part of a group known as the Coelenterates—the word refers to a "hole inside," and their simple cuplike anatomy. The Coelenterates also included the corals, sea anemones, and comb jellies. If you live on the East Coast of the United States, you might be familiar with comb jellies as gemlike clear balls called sea walnuts that sometimes wash up on the beach. In aquaria, comb jellies are admired for the rainbows they throw off in the bright light, which are made by the comblike rows of cilia they use to swim. But more study showed that jellyfish, corals, and sea anemones share many similarities that comb jellies don't, especially when it comes to their stinging cells. Jellyfish, corals, and sea anemones fire poison-laden darts at their prey. The word for these stinging cells is cnidae, which comes from both the Latin and Greek meaning "nettle," because they leave a red welt like the plant with the same name. Comb jellies capture prey with a different type of firing cell. They deploy an adhesive suction cup rather than a pointed dart, which is why comb jellies don't leave welts and you don't need to be worried about swimming into a swarm of them. Today, the Coelenterates no longer exist. The comb jellies have been split into their own phylum, Ctenophora, which means "comb-bearing." And jellyfish, corals, and sea anemones are in the phylum Cnidaria, for their cnidae. Despite their more recently acknowledged evolutionary differences, jellies from the two phyla continue to be lumped together ecologically and to share the orthographic oddness of a silent *c* at the start of their new names.

As the 1800s faded into the 1900s, around the time that the new theory of evolution was gaining traction, the swimming coelenterates had a moment in the spotlight. Some of the most preeminent thinkers of that time studied jellyfish, including, of course, Charles Darwin, but also the founder of Harvard's zoology museum, Louis Agassiz, and celebrated biologist Thomas Huxley. Even royals found jellyfish fascinating. Prince Albert I of Monaco encouraged Charles Richet, who won the Nobel Prize for the discovery of anaphylaxis in 1913, to study the toxin of the Portuguese man o' war. Emperor Hirohito of Japan was caught up in the jelly craze. His hobby was finding and describing new species. These enthusiasts reasoned that gelatinous animals with the simplest bodies were a link between single-celled creatures like protozoans and multicellular us. They saw the gelata as key to understanding the origins of our more complex bodies and turned to the ocean for raw material. Twice as many comb jelly species were identified between 1900 and 1910 than in any other decade before or since. More of a kind of jellyfish called hydrozoa were identified during that same decade than any other time in history.

At the turn of the century, scientific interest in gelatinous animals oozed into popular culture. Naturalist Ernst Haeckel literally had a hand in popularizing jellyfish, creating exquisite Art Deco–inspired drawings of the creatures. A beautiful rendering of jellyfish in red and green from his coffee-table book *Art Forms in Nature* still pops up on Pinterest every Christmas. In perhaps the greatest testimony to their celebrity, Sir Arthur Conan Doyle cast a lion's mane jellyfish as the murderer in a Sherlock Holmes mystery, "The Adventure of the Lion's Mane." Steve Haddock, whose cautionary words about drawing connections between jellyfish and ocean acidification helped to pull me into the world of jellyfish, summed up the turn of the twentieth century as "a golden age of gelata."

In the early twentieth century, the industrial revolution came roaring into marine science. Instead of wading in the shallows with dip nets or scooping jellies in buckets from the sides of sailboats, scientists began to cruise the world in motorized research vessels. They deployed massive nets, dragged them through the water, and reeled them in with mechanized winches. Faster was not better for jellyfish. They did not fare well in these nets, which shredded their delicate gelatinous tissue. Once on board, the treatment was even harsher. Scientists sometimes poured bleach over their nets to dissolve away the remains of jellies and disarm their stingers before sorting their catch. Marine biology became biased toward the animals that could withstand the nets and the bleach. It became the study of durable animals: fish, shrimp, crabs, lobsters, and shelled plankton.

It wasn't until the 1970s that a few intrepid scientists realized that jellies and other fragile animals had been systematically overlooked for half a century. Luckily, technology had morphed again, and new tools were at their disposal to look at the oceans' inhabitants in a different way. Unlike terrestrial ecologists, who can walk or climb into the world of the organisms they are studying, marine scientists can duck into their subjects' world for only as long as they can hold a breath. The availability of scuba in the 1950s meant that scientists could hold their breath a lot longer. They could effectively take short hikes into the sea, where they could observe the animals they studied in the wild. Even with scuba, however, the work was confined mostly to places near shore until a technique called blue-water diving was pioneered by jellyfish scientists, who recognized that there was just as much to learn in the middle of the ocean. In the open water, where there's no seafloor for orientation, these researchers developed a safe diving system of tethers and scouts. The experience is like being in outer space, surrounded in all directions by only the deep, dark blue. You must reel in your visual focus from the boundless shadows to just a few inches in front of your mask. There, very slowly, the alien shapes of gelatinous animals begin to emerge before your eyes.

Physical bodies naturally recognize other physical bodies in a surfaceless space. That's why nets and trash adrift in the ocean are often crawling with life. Before we added so much trash to the sea, jellyfish were one of the biggest surfaces around, and blue-water divers discovered that, rather than isolated individuals, open-ocean jellyfish were entire ecosystems. They saw shrimplike creatures called hyperiid amphipods scuttling around on jellyfish surfaces, burrowing into jellyfish bells, and laying eggs in jellyfish interiors. If the amphipods were hungry, they'd snack on the jelly the way an ant might snack on a leaf. Blue-water divers saw baby lobsters hitching rides on jellyfish bells, like miniature surfers. They saw jellyfish sheltering dozens of fish within their tentacles, like shrubs with birds nesting in their boughs. And in the Arctic, diving birds called murres took aim for jellyfish, not because they preyed on jellies, but because they preyed on the fish that swam with the jellies. Rather than solitary forms, jellyfish were the heart of floating communities. Ecologically, they filled a niche like that of trees in a prairie.

Around the same time, submersibles and remotely operated vehicles (ROVs) began descending deeper into the sea, a mile or more down. These vehicles were outfitted with cameras that captured a gelatinous world radically different from what research-vessel nets had been pulling up for the previous century. Video revealed a menagerie of new gelatinous animals: comb jellies with huge Dumbo-ear lobes; giant medusae that look like deflated balloons; dinner plate–size jellies with tentacles that point up instead of down; endless chains of linked jellies called siphonophores, equipped with colored lures and stinging coils. Living below the depth where sunlight penetrates, the creatures were nearly all bioluminescent; the deep sea was a disco of flickers and sparkles. Like the way birds use songs on land, animals of the deep use flashes to call to mates and frighten predators.

In Puget Sound, Japan, and Ireland, airplane surveys now map the extent of jellyfish blooms. Sound bounces off jellies' edges, so scientists in the Arctic are using sonar to locate jellyfish below the ocean's surface. Whereas nets towed behind ships tear jellies to shreds, towed video cameras capture identifiable images of jellies from Norway to the Gulf of Mexico. Jellyfish are now being outfitted with satellite tags to track them around the oceans and find out where they go when we can't see them.

Biochemical tools, like DNA sequencing and isotope analysis, have provided yet another way to look at gelatinous creatures. Studies of jelly genomes show that jellies we thought were one species are actually a dozen. Not all jellies are created ecologically equal: they have different tastes in prey, different styles of mating, and even different predators. If the 1900s were the age of gelata, today we are on the brink of a jelly renaissance.

Once I started looking, I found jellyfish stories everywhere. I spent hours reading about their shape, how they swim, what they eat, whether they think, how they reproduce, how they sting, how they glow. I set a Google alert on the keyword *jellyfish* to go off at 2:25 each afternoon, giving me a half hour to devour the latest jellyfish gossip just before my kids got off the bus from school. I waited for that e-mail as if it were *People* magazine. Most news reports claimed that jellyfish populations would balloon around the world in the future, that we were in for a jellymageddon. My favorite headline, in a San Jose newspaper, was "Meet Your New Jellyfish Overlords."

I spent hours trying to cram as much jellyfish science into my skull as

I could. I compiled a huge document of facts and references. I printed out reams of scientific articles. One name that I came across again and again was Jennifer Purcell. With a list of publications on gelatinous creatures that stretched back to the 1970s and required two generous scroll-downs of my computer screen to reach the oldest articles, she is one of the field's most knowledgeable and prolific scientists. I asked her for an interview. Jenny was working in Qingdao, China, as a consultant to the Chinese Academy of Sciences. Despite being separated by eleven time zones, we arranged to meet virtually face-to-face using Skype. When she appeared on my computer screen, I saw a no-nonsense woman with matter-of-fact short dark hair and discerning eyes framed by sensible glasses.

Our conversation ranged over a broad swath of jelly topics, like what they eat (a lot more fish than most people realize), how bad the sting of Japanese giant jellies is (could be quite bad), and how a North American comb jelly managed to invade seas worldwide (it's a "formidable predator"). We also talked about the connections between jellyfish and humans. In 2010, some scientists believed that human activities like overfishing, coastal development, and climate change were causing jellyfish blooms. Others believed the changes were a result of natural occurrences, particularly global cycles like the El Niño/La Niña pattern. It was a divide that in some ways mirrored the climate-change schism in the American population. I asked Jenny if she thought changes in jellyfish abundances were due to natural global oscillations or if they were human induced.

"People see things the way they are now. And they think that's how it is and how it's always been. But what we see now is a very, very changed ecosystem. The number of people on Earth has increased really fast. Aquaculture and agriculture have increased really fast. Global warming is real, and yet we keep burning fossil fuels. Our efforts at controlling these things—such as they are—are minuscule. One of the things I learned early on in my scientific career is that people often want an answer—one answer." She went on to say that when it comes to biological questions, one answer is rarely possible. Jellyfish are all affected in their own ways by predators and competitors, whether they live near the shore or in the open ocean, whether they go through a polyp stage or not. "But I'm not going to say jellyfish populations are only controlled by climate or by anthropogenic factors. It's very complicated. . . . I'm sure that there are climate drivers and there are human drivers as well, which are driving the climate."

Then she trained her focus on me through the electronic space between us and said something that's stayed with me for all the years I've been chasing jellyfish. "*If you dare*, that kind of story is . . . a real story."

She'd turned the tables on me. I was now the subject of the interview. And it felt like a charge, an intention. Jellyfish are indeed obscure and neglected by both science and the public. But that in no way means they are inconsequential, or that their story is. The story of jellyfish raises challenging and important questions, questions about the future of our oceans, our climate, and us.

Looking back, I see that jellyfish came to me at a unique moment in my life. I stumbled on them when the haze of sleepless nights—brought on by young kids' cries and the frenzy of cramming a working day into the scant hours of a half-day of preschool—had begun to lift. The public school system provided a blessed seven and half hours of tuition-free time to catch my breath, a break during which I didn't have to choose between squeezing in a trip to the grocery store without a kid in the basket or taking a shower. The jellyfish arrived just when the morning routine stopped ending in either tears or screams, when dinner was no longer a chaos of spills and complaints, when food was eaten and real dinner conversation sometimes even happened. They came when the physical demands of parenting toddlers began to shift to the more mental ones of raising middle schoolers. When the jellyfish came, a sweet spot had opened.

Though I didn't recognize it consciously, something that had been waiting deep inside me had been looking for the chance to climb to the surface and ask if there wasn't something more than granola bars and gogurt. That the answer was yes wasn't surprising, but that the answer was jellyfish certainly was. During my long foray into the world of jellyfish, I would learn that even in Central Texas, the ocean still washed deeply through my soul. Immersing myself in the science of jellyfish nourished a recognition that the fate of ocean, the very cauldron where life began and that still sustains our life on land, is now in our hands. We are responsible for its future, which is also our own—and that of our children.

Polyp

What's Your Agenda?

It was spring break, and my first jellyfish research trip was off to a rough start. We were lost on a dark, narrow Mississippi road. Our family—me; my husband, Keith; my then nine-year-old son, Ben; and my daughter, Isy, who is two years younger—had left Austin in our minivan ten hours earlier. Everyone was hungry. The map in my phone was telling me that the nearest restaurant was located in the middle of a ditch on the side of the road, but we couldn't make out any restaurant in the thick swamp.

Finally, my cell phone found a few bars. I pushed the green button and, hallelujah, the call went though. A sweet southern voice answered with better directions: "Just look for the rocking chair, dahl. Y'all can't miss us." We drove on. And then, a thirty-five-foot-tall rocking chair rose out of the humid inky night, presiding over rows of parked SUVs like the Sphinx. This gigantic homage to southern relaxation was an overgrown advertisement for the furniture store next door to the restaurant we were seeking. Comfort comes in many forms.

The restaurant was mobbed. License plates from every state decorated the rafters, and banners from SEC colleges were thumbtacked to the walls. The air was thick with the oily, briny smell of fried seafood. We ordered as soon as we were seated, and our waiter quickly brought waxed paper–lined baskets of food to our table. "What are these?" Isy asked, gnawing on a hushpuppy flavored with just the right amount of salt and jalapeño.

"Fried fry," Keith answered, wiping his greasy fingers.

"Best fried fry ever," she said, popping another one in her mouth with her still-chubby little-girl hands that always reminded me of sea stars.

"You've never even eaten that before," Ben responded in his ever logical way. But she was full and satisfied, so his jab had little edge.

In much better moods, we piled back into the car underneath the shadow of the giant rocking chair. The rest of the drive to Gulf Shores, Alabama, was unremarkable except for the lights twinkling off the waterways, the lovely half-moon in the sky, and the exquisite quiet of two kids sleeping in the backseat.

I'd planned our family spring break for the Redneck Riviera because I wanted to meet with a jellyfish researcher named Monty Graham, who worked at Dauphin Island Sea Lab near the tourist town of Gulf Shores. Monty has studied how jellyfish move, how jellyfish eat, what eats jellyfish, how various jellyfish are related to one another, where jellyfish live, and what causes jellyfish population size to change. He even published a paper on how jellyfish can be transported from place to place in the aquarium rocks sold at pet stores. Along with some colleagues, Monty organized the first major conference on jellyfish blooms in 2000, and he has emerged as a leader working to understand global patterns of jellyfish blooms.

The day after we arrived, I left Keith and the kids to play in the pool at the condo we'd rented and boarded the ferry to Dauphin Island; the trip took about forty-five minutes. Beneath the flat hull of the boat, Mobile Bay was green and murky, an appearance due largely to its depth, which is about that of an Olympic swimming pool. I scanned the horizon for dolphins but saw nothing except massive oil rigs.

When I first discovered his research, I e-mailed Monty, explaining that I was starting a project on jellyfish and asking if he'd have a conversation with me.

He fired back an e-mail saying he'd be willing to talk, "as long as you are going into this without a set agenda."

Agenda? Well, yeah, I had an agenda. I wanted to understand why these poorly understood creatures were getting attention from the news and scientific media. Were their numbers increasing, as I'd read in the mainstream press? Or was the question still being debated, as I'd read in the scientific literature? I wanted to know how changes in the ocean were affecting jellyfish. I wanted to understand more about jellyfish, period. That was my agenda. I assured Monty I didn't have any pernicious intentions, and he agreed to meet me, but the e-mail exchange was indicative. Our conversation started off awkwardly.

"Is it okay if I record our conversation?" I asked, pulling out a small voice recorder. "I'll just use the recording if I have a question about what we discuss. That way I can go back and check for context and to make sure I've got quotes correct." In my transition from textbook writer to whatever it was I was doing now—call it jellyfish journalism, perhaps—I was still new to the interviewing process and had discovered that having a recording was a nice safety net. My finger hovered above the record button.

"No," Monty said curtly.

I was a little surprised. During the interviews I'd done for magazines, that had never happened. But I tucked my recorder back into my bag and pulled out my notebook and pen. Monty launched into his story about jellyfish. "The perception of jellyfish is that they are negative. They are cast as the villain. Corals are perceived as health and jellies are perceived as sickness."

"But jellyfish are so beautiful, almost angelic," I countered, willing jellyfish to be my hero rather than antihero.

"You would think that jellies would have an element of charisma," Monty said, "but their charisma is still building." When you spend your life trying to write grants to fund your work, charisma matters a lot. The public supports animals with charisma, and funding is easier to come by.

Animals like polar bears and whales have so much of it that biologists refer to them as "charismatic megafauna." Jellyfish—usually small, apparently faceless, and comprised of goo and sting—are a different kind of creature altogether.

"What's of most concern to most people is not actually jellyfish," Monty said. "What's of most concern is how jellyfish affect people. Jellyfish cause a few major types of problems for humans: They clog cooling systems; they disrupt tourism; they impede fisheries." Let's slow down and take on each of those problems on Monty's list of jellyfish ills in turn.

First up, power plants. Both the kind that use nuclear energy and traditional fossil-fuel plants are often placed along coastlines so that they can suck in seawater to cool their machinery. The water is usually pulled in through massive grates that block the inflow of marine creatures. Mushy and flexible like rubber sink stoppers, jellyfish are excellent clog formers. And jellyfish clogs happen frequently. Around the world, jellies have been notorious for gumming up operations. In July 2015 in Israel, in October 2013 in Sweden, in the spring of 2012 near Santa Barbara, and in the summer of 2011 in Scotland, Israel, and Japan, jellyfish slowed or halted operations at power plants. As I followed these sorts of stories through the years, I'd see news videos of plant workers shoveling giant mounds of jellyfish into shipping containers near the water intake gratings. One of the most dramatic jellyfish-induced power outages occurred on December 10, 1999, in the Philippines. The pipes that cool a massive coal-fired power plant on Luzon Island, which provides electricity to 40 million customers, slurped in a swarm of jellyfish. The city went black, and rumors of a government coup spread. Fifty dump trucks of jellyfish were hauled away from the clogged power plant pipes. "In Japan," a senior scientist from that country told me later in my research, "the number-one threat to electricity is earthquakes; number two: jellies."

And it's not just power plants; jellies are indiscriminate cloggers. Arguably, the most embarrassing jellyfish clog took place in 2006 during the maiden voyage of the USS *Ronald Reagan*. The $6 billion nuclear-powered aircraft carrier made its first international port of call at Brisbane harbor in Australia. During a five-day stay, crew aboard the football field–size vessel capable of holding eighty aircraft displayed its military might for the locals by driving the planes around on the flight deck. Just when the boat was set to sail, however, a bloom of humble jellyfish was swept toward the harbor, clogging cooling systems and incapacitating the mighty warship.

Roughly 1,800 desalination plants make freshwater for people living in arid lands around the world. As at power stations, jellyfish sucked in with seawater create crippling clogs. Jellyfish have cut freshwater production in Oman and Israel by a third to a half at different times. In the United States, desalination has gained ground in response to droughts that drag on for years. In California, the largest desalination plant in the hemisphere, a $1 billion project, opened in San Diego County in December 2015. Six other plants are operational, and eight more are planned. While jellyfish haven't yet hit these plants, they could, because they regularly bloom up and down the sun-drenched West Coast.

While energy and water generation are certainly important to all of us, the second problem on Monty's list of troubles with jellies is somehow much more personal. Almost everyone who's spent time at the beach has felt the painful sting of a jellyfish. And while not all jellyfish stings are bad enough to ruin a day at the beach, some are truly terrible. Worldwide, the National Science Foundation estimates that there are 150 million jellyfish stings each year. In 2017, two thousand people were stung in Northern Australia in one weekend during what lifesavers called the worst stinger season since they started keeping track. In 2016, more than a thousand jellyfish stings were reported from beaches near Daytona, Florida, during one week. At least a hundred people were stung off Waikiki Beach, prompting its closure. A year earlier, in 2015, nearly five hundred were stung by the fierce Portuguese man o' war and other jellies on the beaches of South Texas during just one weekend.

Not all beaches are the same when it comes to jellyfish stings. Some have more jellies than others. Lifeguard stations along the sparkling sands of the Costa Brava in Spain report that they treat over a hundred thousand jellyfish stings each year, accounting for about 60 percent of total injuries. And other beaches contend with more virulent jellies. The deadliest jellyfish, *Chironex fleckeri*, ranges from Indonesia down to the popular surfing beaches of Australia's Gold Coast. It is the size of a cantaloupe, with tentacles that stretch ten feet, and its toxin can kill a man in three minutes. A single animal is said to hold enough poison to kill sixty people. While there's uncertainty in reporting, as of 2016, the animal has been responsible for the deaths of at least seventy-seven people in Australia. There, swimmers are encouraged to wear full-body Lycra stinger suits and swim inside netted confines as protection against these fiercest of stingers.

While the beachgoing public might have significant problems with jellyfish, those problems at least occur intermittently. Number three on Monty's list of jellyfish ills is their impact on fishermen—and most of that sting is in their pockets. In 2000, shrimpers in the Gulf of Mexico reported millions of dollars in losses due to jellyfish when unusually large blooms of an invasive species swamped their fishing grounds. In 2005, Japanese fishermen encountered massive blooms of the giant jellyfish. When the animals' stinging cells mixed with the haul in their nets, the catch became inedible. Fishermen lost up to 80 percent of their income, nearly $300 million in total. In the eastern Mediterranean, scientists recently estimated that jellyfish can cut some fishermen's yearly profits by nearly half.

Jellyfish have had a nefarious impact on aquaculture farms too. In 2007, a toxic jellyfish species known as the mauve stinger, which is common to the Mediterranean, turned a hundred thousand salmon at the Northern Salmon Co. in Northern Ireland into corpses almost overnight. The loss was estimated to be more than $2 million. In October 2013, up to 20,000 salmon farmed by aquaculture company Marine Harvest were slain during a mauve stinger bloom off Clare Island in Ireland. In December 2014, jellyfish killed 300,000 salmon at a Loch Duart salmon farm in Scotland. A report from Scotland and the Orkney Islands revealed that of the

4.7 million fish lost from fish farms, nuisance jellyfish accounted for 60 percent between 1999 and 2005.

However, if you really want to understand just how much damage to fisheries jellyfish can cause, two examples are the most telling. And most of us in the West missed both of them.

By 1988, the Black Sea, the watery crossroads of civilization located at the nexus of Europe, the Middle East, and Asia, had experienced long-term degradation of its ecosystem, a mixture of problems including pollution, overfishing, warming waters, and habitat destruction. Centuries of sewage and fertilizer runoff had depleted the deep waters of the Black Sea of oxygen. Soviet industrialization increased tenfold the volume of water without enough oxygen to support animal life. Overfishing left a further diminished and unstable ecosystem. But these factors didn't cause the fisheries crash alone. What they did was set the stage for a successful invasion by a small, gelatinous animal. Not a medusa jellyfish, but a comb jelly, or ctenophore.

It's presumed that in 1982 a ship leaving the Gulf of Mexico took on ballast, pumping in at least one or maybe a few of these comb jellies, then traveled across the Atlantic, across the Mediterranean, through the Bosphorus strait, and across the Black Sea to the Crimean coast. There it docked at port and prepared to take on cargo. It emptied its ballast tanks, releasing that one or maybe several North American comb jellies into the Black Sea. The animal's scientific name is *Mnemiopsis leidyi*; the genus name translates as "one with a memory" because of its ability to return to an upright position when flipped by a wayward wave. Still, given the havoc that *Mnemiopsis* created, it's almost as if it knew exactly what to do when it found itself in a foreign sea—as if it had a memory of the place where it landed halfway around the world.

Once released from the ship's hold, the tiny comb jelly righted itself and found much to like in the Black Sea's familiarly brackish water. It paddled its comb rows a few times, and soon it bumped into a zooplankton prey. It released a spring-loaded, glue-laden filament and grabbed hold of its quarry. The comb jelly's muscles contracted, pulling the tiny creature to its mouth. Soon the comb jelly bumped into another victim, which it similarly engulfed. And then the comb jelly ate another and another.

Before long, the newly arrived *Mnemiopsis* was healthy, fat, and ready to reproduce, and so it did. Unlike jellyfish medusae, which require two individuals, a male and a female, to reproduce, *Mnemiopsis* is a hermaphrodite that can breed offspring on its own, releasing more than ten thousand fertilized eggs a day. And these comb jellies appear to be immune to genetic abnormalities that in other animals intensify with inbreeding. Some scientists think that this self-fertilization is the usual mode of reproduction for comb jellies, that reproduction involving two individuals is the rarity. This makes *Mnemiopsis* as transportable as a fertilized seed. Just add time and food, and a single animal can become an entire swarm, and so it did. Four years later comb jellies were spotted on the northwest coast of the Black Sea, and a year later along the northeast coast.

And then, as if life wasn't already good for these invading comb jellies, it got better. It warmed up. In 1989, the springtime water temperature in the Black Sea was about 5 degrees Fahrenheit warmer than it had been two years earlier. *Mnemiopsis* reproduce faster in warmer temperatures. Suddenly the comb jellies were everywhere—and in enormous numbers. The combined weight of *Mnemiopsis* in the Black Sea was thought to be half a million tons. One scientist working near the Black Sea at the time told me he remembered seeing bulldozers moving mountains of dried jelly husks on the beach.

In 1988, before the population of *Mnemiopsis* exploded, total landings of fish in the Black Sea were nearly 800,000 tons. Just two years later, the catch plummeted to just 66,000 tons. The previously lucrative fisheries collapsed because the comb jelly explosion ravaged the prey the food chain depends on. The Black Sea anchovy fishery saw a 98 percent decline in profits because of *Mnemiopsis.* Nearly two million fishermen lost their jobs. Estimates vary, but including losses to fishing fleets and processing plants, the cost of the collapsed fishery was thought to be $1 billion or more.

Commissions were formed to study the ecosystem, to help decide what, if anything, could or should be done. The most promising idea was introducing a predator to control *Mnemiopsis.* History teaches that using other organisms to rein in invasive species is like performing surgery on the ecosystem. The introduced animal must excise only the creature that's a pest, no other. It doesn't sound easy, and it isn't. We got it right when we introduced wasps to kill off the mealybug that was destroying cassava crops throughout Africa. The lives of millions of people who depend on cassava for food were saved. We got it wrong when we introduced the mongoose into Hawaii to rid it of rats. The small fox-faced carnivore decimated native bird populations, which have never recovered.

When *Mnemiopsis* reached the Black Sea, it had no predators there. But it did have a predator back home off the coast of North America. Perhaps not the first responder you'd consider, *Beroe ovata* looks like a cellophane bag used to wrap up a fancy cookie. Despite its flimsy appearance, this gelatinous animal is maybe even a fiercer predator than *Mnemiopsis.* It comes armed with the ideal combination of attributes to end the reign of *Mnemiopsis*: It can live in the low salinity of the Black Sea and can survive a range of temperatures from polar to tropical. It feeds on ctenophores, and *Mnemiopsis* is one of its favorites. Its mouth is lined with rows of teeth made out of modified cilia, which grab gooey prey and suck it in like a set of geared rollers. Once its prey is ingested, *Beroe* literally zips its mouth closed until its prey is subdued so it cannot escape. Scientists have tried to pull apart those zipped-up lips, but the tissue around the mouth tears before the zipper opens. *Beroe* can also turn itself inside out if irritated, and it is highly bioluminescent. Those last two abilities wouldn't particularly help with the Black Sea's *Mnemiopsis* problem, but all heroes have cool powers that aren't necessarily critical to the task at hand.

While members of the commissions argued about whether *Beroe* was fastidious enough to be purposely planted in the Black Sea, serendipity intervened to take the question out of their hands. In 1997, *Beroe ovata* appeared just about where *Mnemiopsis* was first sighted. Three years later, researchers estimated that *Beroe* had consumed about 6 percent of the total *Mnemiopsis* population each day. The next year, consumption increased to about a quarter of the *Mnemiopsis* population each day, decreasing its abundance twentyfold.

With the pressure off the zooplankton, it rebounded, and so did the fish. In 2000, scientists saw anchovy in their nets for the first time in five years. Populations of sprat, scad, and mullet also rose. Fishermen started to catch fish again. While not yet recovered to a pre-*Mnemiopsis* state, the Black Sea again has a fishing industry.

Nevertheless, the invasive nature of *Mnemiopsis* is nowhere near sated. The comb jellies have been making their way into neighboring seas: the Sea of Azov in 1988, the Sea of Marmara and the Aegean and Mediterranean two years later. By 1999, *Mnemiopsis* was in the Caspian Sea. In 2006, it was transported farther to the west, reaching the North Sea, the Baltic, and eventually the eastern Atlantic Ocean. The comb jelly has been found in coastal waters off Syria, and in 2009, it bloomed in large numbers off Israel. There are reports of *Mnemiopsis* from the Indian Ocean and from Australia. *Mnemiopsis*, an animal that once clung to the eastern coast of the Americas, is now a global creature. Whether or not it will find conditions that allow it to bloom as it did in the Black Sea again is unknown. Whether or not its trusty predator, *Beroe*, will follow along on its global trek is also unknown.

The *Mnemiopsis* invasion in the Black Sea is not just a story from the past. In the summer of 2016, the comb jellies reached abundances never before seen in the Adriatic Sea between Italy and Croatia. A column of ocean one square meter at the surface extending to the seafloor contained 500 individuals. This population seems to have different genetics from those of the Black Sea invader. Scientists suspect that the Adriatic bloom was a repeat invasion, suggesting that ballast water continues to transport comb jellies from other parts of the world. The same year, a project doubling the size of the Panama Canal was completed. In 2013, only three ports along the U.S. coast were equipped to receive the supersize ships built for the canal's New Panamax size and draft limits, but plans are under way for renovations from Florida to New York to accommodate the behemoth boats. Another, larger, canal is proposed to slice through Nicaragua, further connecting the Atlantic and Pacific at their midsections. In 2015, the Suez Canal was expanded, doubling the ship traffic between the Indian Ocean and the Mediterranean. It's impossible to guess what creatures will inhabit the increased volume of ballast water passing between the world's oceans. We can't predict which creatures will find themselves in the right salinity and temperature, with plenty to eat and without their native predators. We can't forecast how disruptive they will be to ecosystems and economies once they arrive. We can't know which animals will find themselves in a new place where their old memories might make sense. But some of them will be jellyfish, because they have already proved they can do it. Globalization, the process by which differences become smeared out via exchange, is not confined to our economy or our culture; it's arriving as gelatinous animals in the ballast water of ships.

The second major jellyfish crisis is still going on today. It's been unfolding in a part of the world that often gets overlooked in our Western newspapers: off Namibia. Before the 1960s, these seas were among the most productive in the world. Today, the weight of jellyfish in Namibian waters is more than three times greater than that of fish.

Situated just north of South Africa on the Atlantic side of Africa, Namibia's western coastline is dominated by ancient ochre dunes often blanketed in thick fog that condenses when the cold waters of the Benguela Current, which flows northward from Antarctica, hit the hot desert air. In places like Namibia, where the winds race along the western coast, currents peel back the surface waters, letting fertilizer-rich waters from the deep ocean rise to the surface. And that's why the well-fertilized waters off Namibia grow some of the lushest phytoplankton in the world. Where there's food, there are fish, and Namibian water historically boasted some of the best-known fisheries. Prior to 1987, the harvest of sardines and anchovies was never less than one million tons per year.

During the second half of the twentieth century, Namibia was occupied by South Africa, and fishing its seas was like the Wild West's gold rush. South Africa didn't bother to keep track of who was taking what. From many nations, massive ships equipped as fish-processing plants set up shop offshore, pulling up tons of gleaming hake, horse mackerel, sardines, and anchovy. It was a fishing free-for-all.

A warning came in 1963, when there was a small crash in the fisheries' catch, likely due to an El Niño. When the climate shifted back, fish yields recovered, and the rush was back on. But during that blip, two jellyfish made their first appearance in Namibian waters. One was a more conspicuous creature called the compass jellyfish, *Chrysaora fulgida*, with a diameter of about ten inches. The other was a smaller animal, just four inches wide, called *Aequorea forskalea*.

In any gold rush, eventually the gold runs out. In 1990, when Namibia received its independence from South Africa, fish landings had dwindled. The overall fish catch fell short of 100,000 tons, just a tenth of pre-1987 hauls. The sardine catch came down to just 2 percent of what it was in the 1960s. During a 2003 research survey, the mass of jellyfish didn't just equal that of the major commercial fish (horse mackerel, hake, and small fish like sardine and anchovy). The jellies outweighed the fish 2.4 to 1. Today, if you put your nets in the sea off Namibia, they will likely be bursting with medusae. In these waters, which were once among the richest in the world, seabirds and seals are starving.

While overfishing was clearly a culprit, there's still speculation about exactly what happened. Were the jellyfish able to find a foothold in Namibian waters in 1963 because their fish competitors had taken a hit from the El Niño? Or was it the constant pressure of fishermen culling the fish from the water that made way for the jellyfish? Did trawling change the seafloor, making it more hospitable to the part of the jellyfish life cycle called the polyp?

Regardless of the reason, the prognosis isn't good for Namibian fishermen. A back-of-the-envelope calculation by scientists studying the problem showed that under current conditions, which are warmer and richer in zooplankton than in the past, a single compass jellyfish polyp could easily produce 2 million medusae per year, suggesting that a flip back to a fish-dominated ecosystem is unlikely. The scientists who have documented the changes in Namibian waters warn, "Jellyfish play potentially major controlling roles in marine ecosystems and, in this era of apparent jellyfish ascendancy, marine ecosystem managers and modellers cannot afford to ignore them."

Neither can we. The Black Sea and Namibia are case studies showing that our worst fears can happen. Jellyfish do have the capacity to take over ecosystems, creating a sea full of goo. But does it have to be that way? Is it inevitable?

In a 2011 study, Jenny Purcell, who dared me to write a book that mattered, combed the scientific literature for data on jellyfish-based problems. One of the few places she found good data was Japan. She looked at the three categories on Monty's list of jellyfish ills: clogging power stations, stings, and fisheries. From the 1920s to the 1960s, records of negative impacts by jellyfish were confined to fisheries. Beginning in the 1960s, reports appeared of jellyfish clogging power station intakes and causing significant numbers of stings at beaches. In the most recent decade of her study, 2000 to 2010, incidents in all three categories more than doubled.

Common sense partially explains why human-jellyfish interactions are on the rise. In the hundred years between 1800 and 1900, Earth's population increased by about 600 million. In the next hundred, between 1900 and 2001, it increased by nearly 5 billion. In the twelve years between 2001 and 2013, we added another 2 billion. Just as important as the exponential human population growth is the fact that roughly half of our ever-expanding populace lives along the coasts. In the continental United States, the coasts make up only about 20 percent of the land, but more than half of the country's inhabitants live there. Worldwide, 75 percent of all large cities are coastal. One reason more people are having problems with jellyfish is that there are more of us around the coasts to run into them. And that makes it all the more confusing to understand jellyfish. "It seems clear that human problems with jellyfish have increased recently," Jenny wrote. "Whether jelly populations per se have increased is less clear."

Perception is reality," Monty told me several times while we talked in his office, and it really is when it comes to jellyfish. Jellyfish populations are by nature patchy in space and variable in time, but we don't realize it. In most waters at most times, jellyfish medusae are represented by just a few individuals; they're just another part of the oceanic medley. But at other times—times that we are still unable to predict—jellyfish have a mind-blowing ability to rapidly increase their numbers and create blooms. To some extent, those blooms ride the ocean at the fancy of tides, waves, and currents. This blooming and sloshing through coves and gyres means that, in any one spot, jellyfish numbers can change dramatically over the course of hours, days, and weeks. It means that they are unpredictable over the course of months and years. This is maddening for scientists trying to reserve ship time years ahead to study jellyfish.

Jellyfish and human perception work on fundamentally different time and space scales. We remember things when we encounter them. If we see a lot of jellyfish littering the beach during spring break—or worse, if we get stung—and that didn't happen the year before, we draw the conclusion that jellyfish are on the rise. Whereas a lack of jellyfish encounters is a non-event. It doesn't lodge in our memories. To make things worse, our memory fades with time. Was it last year we saw jellyfish on the beach, or the year before? Our recollection becomes fuzzy over decades, and impossible over generations.

In 2010, the perception that jellyfish numbers might be on the rise around the world inspired a group of about two dozen scientists to collect all the existing studies on jellyfish, spotty as they may be, and plumb them to see whether the data showed that jellyfish populations were really increasing. Led in part by Monty, this group won a three-year grant with the National Center for Ecological Analysis and Synthesis, or NCEAS, headquartered in Santa Barbara.

Monty told me, "NCEAS is trying to generate the tools to answer the question: Are jellyfish populations increasing or decreasing? And I think we should have an answer in a couple years." Monty leaned back in his chair and sighed. "At one point, I was really concerned with getting that answer. We thought we'd get a big database and run some algorithms and get a solution. But in the last couple months I had a moment of clarity. So fucking what?"

Wait. What? Did he really just say that? I felt as if I had lost the thread of the conversation. But before I could understand what he meant, Monty was ushering me to the door. My time for asking questions was over. I was dismissed. And I was much more confused than when I had walked in.

Across the hall from Monty's office, I met Kelly Robinson, a PhD student who had been tasked with assembling the world's largest database of jellyfish information for the NCEAS project. The group cleverly named it the Jellyfish Ecological Database Initiative, or JEDI.

Kelly showed me how the work was progressing. She had collected a quarter million records from around the world and was in the process of standardizing them so they could be analyzed in the same database. When she flipped her computer monitor from rows of data to a map of the world, the most obvious thing I saw was that the coverage was very uneven. There was data from the coasts of the United States, including Alaska, and throughout Europe, but very little from Africa and the Middle East. That was to be expected, Kelly said. Collecting jellyfish data has always been an intermittent practice.

A jocular Australian stopped by Kelly's lab. His name was Rob Condon, and he also worked on jellyfish ecology and was a member of NCEAS. Rob asked me to join him, Kelly, and a jellyfish geneticist, Keith Bayha, for lunch. Over fried clams, Rob said that he had posted a poll on Facebook asking jellyfish scientists to chime in on whether or not jellyfish numbers were increasing globally. When I checked later, I discovered that the count was much more ambivalent than the news articles I'd been reading. The vote stood at 7 maybes, 2 yeses, 1 no. Rob had told me he had cast the no vote; he didn't believe that there was enough evidence to show that jellyfish were taking over the oceans. There was one more vote, a write-in, I guess. It read, "don't care." That, of course, was Monty's.

The next day, our family went to the beach. We parked on the sandy shoulder and looked across the sand. Keith asked, "Does the beach look closed?"

Huge construction equipment, yellow dump trucks, jeeps, and trailers lumbered or were parked among the dunes. Just beyond the trucks, the pastel colors of girls' swimsuits and the black sun-protection shirts of small boys stood out against the surf. Shade tents were secured in the sand. The beach looked open but under major renovation.

We gathered our chairs, toys, water bottles, and sunscreen. About halfway to the water, three-foot-tall stakes marked out a sort of road in the sand. Construction equipment rumbled by. We looked left and right, just as we would at a real street, and then scurried past. We set up camp on the beach, and the family sitting to our left approached, offering to share beach toys. They were from Dallas. As a monster truck hauling a wire mesh scoop drove past, they told us they'd been here before.

"What's going on?" I asked, gesturing to the equipment.

"It's cleanup from the oil spill. They're doing a great job."

It had been nearly a year since the Deepwater Horizon oil rig exploded in the Gulf of Mexico. Almost three months passed before the gusher could be capped, making it the largest accidental marine oil spill in history, a release of over 160 million gallons of petroleum into the Gulf of Mexico. BP would end up paying more than $60 billion in damages. As I looked around, I thought that the beach did look clean, not a speck of trash anywhere. We played Frisbee, though the only one of us who was coordinated enough to catch it reliably was Keith. Ben, his nose growing ever more freckled in the sun, built complex roads for his beach toys in the sand, which Isy tried to decorate with stray seaweed and shells. He told her to quit it, so she ran off to dance and leap over waves, much more daring in the water than her older brother would ever be. We watched seagulls and ate pretzels whose crunch was crunchier because of the sand. Was it sand that had been recently cleaned of oil? I wondered as I licked my fingers. It was so pleasant that thoughts of petrochemical residue mixed with the warming rays of sun and evaporated into the cloudless sky.

After a while, I noticed two men wearing orange construction vests walking the beach. They carried fishing nets and had duct-taped the tops of their yellow booties to their jeans. I walked over and asked if they were scooping up oil in those nets.

The older man, who looked to be in his fifties, had only about half his teeth. He pulled out a gray pebble from his net and pushed on it with his thumb. It broke into two pieces. Oil disguised as a rock. "They pick up about a thousand pounds of this stuff a day and send it to a landfill somewhere in central Alabama."

I asked if the washed-up oil could be used for anything. The older man said that it hadn't been cost-effective. The younger guy chimed in, long bangs of a bowl cut nearly covering his eyes. "I heard they were going to figure out a way to use it for asphalt."

When I asked how much longer they'd be doing this work, the older man told me, "About two more years. They say the bottom of the Gulf is two feet deep in this stuff. The tide brings it in. It's going to be bringing it in for a long time. Last summer when the spill happened, this beach was two feet thick in the stuff. We had to use shovels back then. Some of the oil was this big across." The older man stretched his hands as if holding a basketball.

Thoughts of petrochemical residue on sandy pretzels condensed, formed droplets, and rained back into my mind. I thought of the fresh fried delights I'd eaten the night before at Catfish Charlie's. "How's the seafood?"

"They say it's fine. I eat it all the time. I personally think you're going to have to wait a couple years to know. Until it makes its way up the food chain. But since there's so much down there, it's gotta go somewhere." The man smiled a knowing smile, revealing dark spaces between his remaining teeth.

I didn't know it then, but work was already under way studying the unlikely connection between jellyfish and oil. While 1.7 million gallons of chemical dispersant were used to break up the slick and sink it to the bottom of the sea, scientists were learning that humble jellyfish might act as a natural, and safer, way to clean up an oil spill. The goal of spraying dispersants on the ocean is to break apart the oil into tiny droplets. Getting that breakage to happen requires mixing. When thousands of jellyfish swim, they mix the water a lot, maybe enough to break apart oil slicks without adding so many expensive and potentially harmful chemicals. And jellyfish make a lot of mucus. Like a sticky net, the mucus collects bits of oil and detritus. The conglomerated mass is more likely to sink away from the ocean surface. The mucus also contains a lot of nitrogen, which important oil-consuming bacteria crave and which is often missing from ocean water. The extra nitrogen stimulates the growth of those helpful bacteria. While jellyfish do cause a whole host of problems, we are just beginning to understand the ways they help correct others.

On our last day on the beach in Alabama, as one dad helped his son cast a fishing line into the surf, he saw a jellyfish washed up on the sand. I ran over to take pictures. Though my knowledge of jellies was still flimsy, I knew this species. It is the most widespread jellyfish in the world, the moon jelly, *Aurelia*. This one was about four inches across, whitish but translucent, with a characteristic four-leaf clover design on its bell.

Down the beach, the oil-cleaning operations continued. A giant shovel piled sand into a gargantuan dump truck. The sand made its way to a blue conveyor angled to the ground and reaching as high as a two-story building. Sand was dribbled onto the bottom of the conveyor and was shaken through a grating on its way back up. Pieces of oil gravel remained on the grating and spilled over the top into another monster-size shovel, apparently headed for that landfill somewhere in central Alabama. The entire beach was being dug up, sifted, and poured back in place by huge construction equipment.

A couple of kids came racing over. "We found a giant jellyfish!" This one was so big, it had to be carried in a wagon. A few kids dragged it up the beach and dumped it on the sand. It was a massive jelly, three feet across. We all touched its top surface. The kids turned it over. Someone stuck a sand shovel through the outer bell, splaying open clear jelly innards. No one could resist. In a frenzy of pink, yellow, and green beach toys, the jellyfish was cut to shreds.

One kid exclaimed, "I don't care about this stupid jellyfish."

"It's not stupid," Isy said, defending me and my new fascination.

"Yes it is. It doesn't have a brain," the first kid responded. He wasn't quite right. I would learn that whether jellyfish have brains depends on your definition of a brain and that jellyfish have a lot of ways of navigating their world that require complex, albeit distributed, processing. But neither Isy nor I knew that then.

Isy hesitated a moment, picked up a shovel, and joined the chopping. When it was over, the kids bolted away and left me staring down at the remains of the jellyfish, a scrambled mess of sand and goo. It was tempting, after talking to Monty, to think that the kids' destruction of the carcass revealed some instinct to fell the villain, to desecrate the evil that's interfering with our power plants, our food supply, and our vacations.

But that's not true. The truth is that jellyfish are just doing what they have done for eons, and the kids didn't see the jellyfish as a symbol of harm. Rather, they were simply acting on their impulses. The first thing they—and I—wanted to do was touch the jellyfish, to know what it was made of, to explore the inside of this alien creature. Cutting into the jellyfish felt good. And as with my research into jellies, once started, it was impossible to stop.

3

Jellyfish Salad

A few weeks after we returned from Alabama, I stood in a parking lot bordered by strip malls on the northern edge of Austin. The flat-roofed buildings housed several Vietnamese restaurants, a Target, a used-furniture store, a liquor store, and Hong Kong Super Market, its flashing neon OPEN sign beckoning me. If I was going to bite off the task of writing a book about jellyfish, I'd decided I'd better take a bite of an actual jellyfish too. I headed into Hong Kong Super Market hoping that it carried jellyfish, which has been eaten in Asia for over a thousand years.

I wandered around aimlessly at first, hoping to stumble upon some sort of highly processed jellyfish, maybe even coated in a peppery seasoning and sold in a familiar foil snack food package to lend anonymity. In the frozen section, I saw croaker, grouper, brick-colored shrimp the size of fingernails, ruffled slices of what looked like abalone, tubs of some sort of fish paste, puny squid the size of a half dollar and sizable specimens about as big as a slipper. I strolled into the canned foods aisle. I saw something that looked tentacle-ish on a package, but it was dried squid chips in two different intensities: hot and mild.

One of the checkout lanes was empty, so I asked the saleslady, "Do you sell jellyfish?"

"Hmmm, we may be out. Let me check." I was surprised that she wasn't surprised. I followed her through the aisles of cans and around the shelves full of rice cakes, toward the refrigerated section. There, among the tubs of udon noodles, she located what I'd been seeking: a clear plastic bag with a white butcher's label that said "Yelly Fish $4.26." Inside was a large taupe-colored lump.

I looked curiously at the lump. What was I going to do with that fleshy clod?

The bag fell heavily into my hands. "How do you cook this?" It was weightier than I expected.

"What? You don't know how to cook? Oh, you have to wash many times."

"Wash?"

"Yes, very salty. You must wash salt away."

"And then, do you boil them?"

"Yes, you dip in hot water."

"What do you do after you boil them?"

"You must slice."

"And then?"

"Then eat in salad. Maybe with vegetables. Maybe with shrimp."

I followed her to the front of the store and made my purchase. She dropped the jellyfish in a plastic shopping bag. One that just might find its way into the ocean and be mistaken for a jellyfish itself someday. But momentarily I wasn't thinking of that. I was contemplating just how one goes about washing a jellyfish.

The checkout lady also seemed concerned about my plan. She called after me, "You must wash, many times."

I placed the bag of jellyfish on the passenger seat of my minivan as if it were a dead body, which it was. Several, really. But despite the feeling of trepidation I had about the alien flesh sitting next to me, I found one thing marvelously surprising: It's very easy to buy jellyfish in landlocked Austin, Texas.

It took me several hours to muster the courage to pull the bag of jellyfish from my refrigerator and open it. When I did, the smell wafted up: fishy. I ripped open the outer bag and saw a second layer of plastic. The smell intensified: distasteful. I dumped the whole mess into a colander. The jellyfish looked like layers of wet cardboard wadded up inside one another. I pulled one out and held it up. It was a rounded disk about eight inches in diameter with jagged ruffles around the edge. This was the bell of a jellyfish. It felt wet but not at all slimy. It was tough and kind of leathery; it didn't feel fragile at all. In the light, the color was a dark tan, and where it was thin, it was translucent like amber.

While jellies aren't found on Western menus today, we once considered them delicious or at least palatable. In *The History of Animals*, Aristotle discussed the culinary aspects of jellies:

> Of sea-nettles there are two species, the lesser and more edible, and the large hard ones, such as are found in the neighborhood of Chalcis. In wintertime their flesh is firm, and accordingly they are sought after as articles of food, but in summer weather they are worthless, for they become thin and watery, and if you catch at them they break at once into bits.

Two millennia later, in his seminal work on jellyfish, *Medusae of the World*, published in 1910, Alfred Goldsborough Mayer wrote about eating *Rhopilema esculentum*, the most popular culinary jellyfish: "This medusa is abundant in the Inland Sea of Japan, and it is also found off the coast of China. It is the custom in Japan to preserve it with alum and salt, or between the steamed leaves of a kind of oak. It is then soaked in water, flavored with condiments, and when so prepared constitutes an agreeable food."

On the other hand, the 1996 manual *Venomous and Poisonous Marine*

Animals: A Medical and Biological Handbook reported: "*Rhopilema hispidum* is on sale in some restaurants in the West End of London as a 'delicacy.' The taste was described recently . . . as 'dreadful.'"

Would I be cooking something agreeable or dreadful? I didn't know. But I did know that the jellyfish in my colander were none of the species that I'd been reading about. I was pretty sure this was a cannonball jellyfish (*Stomolophus meleagris*), also called a jellyball, or locally on the Texas Gulf Coast, a cabbage head. The animals are named for their high-domed bell, which is cream-colored with an edging of brown speckles. They have short oral arms—the frilly extensions that hang off their mouth like exaggerated lips. They also lack tentacles and are only mild stingers. The jellyfish I had purchased could have hailed from Baja Mexico, where a new cannonball jellyfishery was on the rise, or they could have been from the Gulf of Mexico, where about two dozen fishermen target jellyballs. Maybe they were caught off the coast of Georgia, which meant they would have been caught by one of the five or six licensed jellyfish fishermen in the state.

One of those Georgia jellyfishermen was Wynn Gale, nicknamed "the jellyball king." His Facebook cover photo doesn't show him or his family, but the cartoon character SpongeBob SquarePants, who also relishes the art of jellyfishing. Wynn's posts include photos of the jellyfishing boat he captains, the *Big Cobb*, bearing the flags of Georgia and the Confederacy, so weighed down with jellies that its fantail sits just a few inches above the water. Videos he's posted show a net so packed with jellyfish, it takes the shape of a giant teardrop. It jostles and jiggles, the weight of 6,500 pounds of jellies straining against the netting. Wynn maneuvers the catch over the back of the boat with a winch and then a crew member pulls the release cord, opening the bottom of the net. Jellyballs cascade out until the back of the boat is covered shin deep. The catch is swept into the hold just as the next net filled with another 6,500 pounds of jellyballs appears over the opposite side of the boat.

I caught up with Wynn one day in April after jellyfish season had ended. He was working on the jellyfishing boat, changing the nets back to smaller-gauge shrimp nets. I asked him about his jellyball season, which he told me was short but successful. When the jellies are abundant, he said, he could fill his nets in about fifteen minutes. He'd usually call it a day when he'd caught about 60,000 pounds, but one time, when the water was calm and he was fairly close to shore, he maxed out at 78,000. During jellyfish season, which happens in between the white and brown shrimp seasons, in the late fall and winter, Wynn might take three trips a week.

The jellyfish processor pays Wynn 6 or 7 cents a pound, so a full boat is worth about $4,000. After splitting the money with the boat's owner and paying for gas and the crew, for every million pounds of jellyballs, Wynn still makes $10,000, which is the major reason he likes jellyballing so much. The money is good. And compared with other fishing, jellyfishing is a clean fishery, he said. The big holes in the nets minimize the bycatch, the fish he catches inadvertently and throws back to the sea. The nonprofit Oceana has identified bycatch as one of the major threats to healthy fish stocks worldwide. By their estimate, 63 billion pounds of bycatch are captured annually, equivalent to 40 percent of the fish removed from the sea. A study of Georgia jellyballers found that their bycatch was nominal, especially compared with shrimping.

But Wynn loves jellyfish mostly for the sake of jellyfishing. "Jellyfishing is hard, fast fishing," he said. He has to bring the nets up fast, and get the jellyfish in the hold fast, and get back to the processing plant fast, otherwise the jellyfish start to sour. He likes catching the big blooms of jellies when he can fill his nets in minutes. He likes figuring out where the jelly blooms are.

Unlike fish, which have gas bladders that sonar bounces off, jellyfish are mostly water and invisible to fish finders, which means Wynn has to know how jellyfish behave. Jellyfishing is best on calm days. "That's when they'll be just about ten feet below surface, where you can catch them. When the wind comes up or the barometric pressure drops, they'll dive down in unison. Within ten minutes they'll go straight down and you ain't gonna find them. It'll make you scratch your head." Compared with shrimping, which Wynn finds monotonous, jellyfishing keeps him on his toes and it's a lot more fun. "Jellyball fishing is crazy. It's a sickness. Sometimes, it's all I can think about." A British production company recently discovered jellyfishing and is planning to develop a reality TV show around Wynn and the other jellyfishermen in Georgia. "They are going to call the show *Big Balling*," Wynn said.

Wynn would catch jellyfish every day of the season if he could, but the jellyfish processor holds him back. In order for a fisherman to get a jellyfish permit in Georgia, he has to prove there's a processor willing to take all he catches. There's only one licensed processor in Georgia, called Golden Island International. Before the season starts, Golden Island issues a letter to the state saying how many jellies they will buy based on what they think they can sell to their markets in Asia. The fisherman can't exceed that limit. But the catch keeps going up. In Georgia, jellyfishing is now the third-largest fishery by weight, about to eclipse crabbing. Besides Georgia, a handful of boats jellyfish the waters of Florida as well. By some estimates, the Southeast's jellyfishing industry is worth somewhere in the low millions and growing.

The Golden Island International jellyfish processing plant is managed by April Harper. April hadn't ever thought about jellyfish before her husband, who worked at a bank in Darien, Georgia, mentioned that one of his customers was looking for administrative help. "Jellyfish would have been the last thing on my mind," April told me. But she was bored at her government job and went to work for the client, Terry Chuang, who is from Taiwan and well versed in the secrets of jellyfish processing. Over the years, Chuang shifted more responsibility to April, and now she oversees much of the operation.

When a boat like Wynn's pulls up at the dock, a hundred people work for six to eight hours to unload and process the tons of catch. The first step is to vacuum the jellies from the ship into slotted tanks that can hold eight hundred pounds each. Water immediately begins to drain out of the jellies through the slots. And getting the water out of the jellies is the whole point of jellyfish processing. Jellyfish release a sugar-rich mucus that bacteria love, and liquidy jellies can quickly rot. The second step is to separate the oral arms from the bell, an operation done by hand. "It's very difficult," April said. "You need a good clean cut. We tried to develop a machine to do it, but we haven't found a piece of machinery that has the right qualifications yet." After the parts are separated, the jellies then go through a washing process in what April described as "an agitator like an old washing machine" for about twenty minutes to continue to get rid of the water and slime.

Next, the jellies go into their first salt bath, which is 200 to 250 pounds of salt and 1 percent alum, or hydrated potassium aluminum sulfate. Alum is a translucent solid that has been used for thousands of years for pharmaceutical and culinary purposes. The Romans used it to purify drinking water, and it's used in some deodorants. Recipes for maraschino cherries and pickles sometimes use alum because it makes the flesh firm and crisp, and those are qualities that jellyfish processors are looking for as well. Ideally, prepared jellyfish has a crunch similar to a carrot or celery, April told me. "The key to a good product is to use the right amount of alum. If you use too much alum you get a brittle product. Low alum makes a good, viable, flexible product."

Using alum requires care not just because it affects the crunch. Alum can be unhealthy. It has been linked with both dementia and Alzheimer's disease. The woman from Hong Kong Super Market told me to soak the jellyfish many times in order to remove the alum. April told me about a study released in China in 2014 that found people were eating unsafe levels of alum, and one big reason was jellyfish. A dotted line was drawn to suppliers from outside Asia with weak quality control. Sometimes when the jellyfish fishing is good, fly-by-night processors set up shop on the beach. The rumors were that these processors didn't have rigorous protocols and their jellyfish were tainted with too much alum. Demand fell, buyers were scared off, and the market for jellyballs tanked. In response,

Golden Island developed rigorous safety standards at its facility, April said, working with government agencies and universities to standardize processing methods.

After the jellyfish soak in the first salt bath for about a day, they are moved to a secondary salting bath, where they remain for a week to ten days. The salinity in this second dousing is constantly monitored to ensure that it stays high enough to prevent bacterial growth. The jellyfish are then drained. By this point they look like the leathery tortilla that I pulled from the plastic bag I bought at Hong Kong Super Market. They are put in five-gallon buckets, packed up, and sent to China or Japan or to Asian supermarkets in the United States. In this condition, the jellyfish product has a shelf life of two years at temperatures below 70 degrees Fahrenheit. They are essentially jellyfish pickles.

"It's not a finished product, though," April cautioned. Just like the woman at Hong Kong Super Market, she told me, "You have to rehydrate it. Let it soak in cold water for eight to ten hours. You have to make sure you get the salt out."

Back in my kitchen, as I stared down at my jellyfish, I heard April's voice and the checkout lady's words in my head. I rinsed, and rinsed, and rinsed the jellyfish. Eventually, the heavy fishy smell dissolved. I put the washed jellies in a pot, covered them with more fresh water, wished them well, and put them to bed for the night.

The next morning, I changed my jellies' water again. They seemed to have puffed up a bit. I could see some of the details of their bell, some pointed fins, the circle where the digestive system or maybe gonads attached to its center. The color had lightened slightly.

None of my cookbooks had jellyfish recipes, so I turned to the Internet for guidance and found more than I could ever need. A blog called *Deep End Dining* introduced a step-by-step jellyfish cooking procedure with a sci-fi play in which jellyfish had taken over the world and were haunting a certain ex-president for his less-than-stellar environmental policies. I put my trust in *Deep End.*

Deep End suggested rolling the bell up like a sleeping bag before slicing it, which I did. Generally, I don't like cutting meat. I don't like trimming chicken fat, beef fat, or pork fat. I find skinning fish difficult. The rubbery texture, the way that the meat does not seem to want to be separated from itself, disturbs me. By comparison, cutting jellyfish was enjoyable. The knife passed through the rolled-up jellyfish easily, but not too easily. There was a firmness, but not the kind of resistance you get with meat. It was fun, kind of like cutting sheets of pasta.

I dumped what was now jellyfish linguine into a pot. *Deep End* told me to pour boiling water over the sliced jellyfish, stirring with a chopstick occasionally. I poured. The jellyfish, for the first time, looked like animals. Wormlike, they wriggled and twisted in the boiling water. The rough edges of the bell quaked, the small fins flicked, the thicker ridges waved. I stirred. The jellyfish looked like linguine again.

Deep End's marinade included white rice vinegar, soy sauce, sesame oil, and sugar, which I prepared. I tossed the jellyfish and put the salad in the refrigerator to marinate. I still wasn't sure I wanted to eat these jellyfish. I was fine to wait for the flavor to set.

A serving of jellyfish contains just 25 calories but a lot of protein, 6 grams. If you concocted a 200-calorie pure jellyfish energy bar, it would contain 48 grams of protein and no fat. For comparison, an average 200-calorie protein energy bar, developed in test kitchens with the express function of delivering a wallop of protein, would have less than half as much protein, at least 5 grams of fat, and several grams of sugar.

Jellyfish are so full of protein because they are largely collagen, which is the most abundant protein in our own bodies too. It is a complicated molecule made of three chains that twist together in a triple helix, which then further combines with other triple helices to form long, strong, stretchy fibers. Those fibers are used to build our skin, tendons, veins, organs, bones, cartilage, cell surfaces, and hair. In medicine, collagen is used to promote wound healing, for cosmetic enhancements, and even as scaffolding on which to grow synthetic organs. Today, much of the collagen for medical uses comes from chickens or pigs, and it can cause allergic responses in some people. While still speculative, scientists are investigating jellyfish protein as an alternative, and two companies in Europe have begun harvesting jellyfish in order to extract the collagen for pharmaceutical applications and biotechnical research.

In studies done in China, scientists shaved a patch of skin on the backs of mice and exposed them to ultraviolet light. Some were fed jellyfish collagen; others weren't. The incidence of skin damage was significantly lower in those that ate jellyfish. And in a study in the United States, collagen from the cannonball jellyfish, which was currently marinating in my refrigerator, suppressed rheumatoid arthritis in rats. These results are thought to be related to the ability of jellyfish collagen to act as an antioxidant.

Antonella Leone, a scientist at the National Research Council's Institute of Sciences of Food Production in Italy, analyzed three species of jellyfish for their biological activity. Remarkably, the antioxidant activity of jellyfish collagen was higher than collagen collected from chicken and used as a standard measure. She also found that extracts from the jellyfish contained high amounts of omega-3 fatty acids, factors that can be deficient in the Western diet because of its reliance on processed fats but which remain essential because of their antioxidant activities and anti-inflammatory properties.

When I spoke to her on the phone, Antonella explained that the apparent health benefits of jellyfish coupled with the potential that jellyfish could be fished sustainably make jellies an exciting novelty for discriminating Italian foodies. The fact that Italy has a large untapped resource in jellyfish makes the idea attractive to fishermen. "We really have a large biomass of jellyfish in the Mediterranean," she told me. "Right now the Ionian Sea is full of *Rhizostoma pulmo* [barrel jellyfish], which can reach forty to fifty centimeters [about a foot and a half] and weigh several kilograms [six to eight pounds], and they are completely edible."

In 2015, Antonella organized an event at the Milan Expo, a World's Fair–like festival. Called Novel Foods, the gathering brought together experts on different types of foods that make sense to convert from novelty to crop. Headliners were crickets, algae, and jellyfish. Antonella asked a renowned Italian chef, Gennaro Esposito, to create jellyfish dishes. "People were very impressed and seemed ready to accept this kind of food," she said. But she noted that Europeans object to undermining the health benefits of jellyfish by the use of alum, so she's been studying new processing technologies. Ideas like high pressure and extreme cold look promising, but more work needs to be done.

"After tasting all the jellyfish at the Expo, what was your favorite way to eat them?" I asked Antonella.

"Fried." She laughed. "Really. Jellyfish have no taste. Just a crunchy composition. I enjoy it as salad and as tempura. Chef Esposito prepared it with a salad seasoned with salt and sugar, and it was very good. He asked if I had more jellyfish for him because he wanted to include it on the menu in his restaurant."

The night that I put jellyfish salad on the dinner menu, only Isy and I were eating at home. Looking back, I do think it's a bit suspicious that both Keith and Ben managed to be elsewhere. After a few hours in the refrigerator, my jellyfish salad had taken on a deeper brown color, probably from the soy sauce marinade. I placed a small pile on my plate. I offered some to Isy, usually a brave eater, but she politely refused. My Internet recipe suggested a garnish of sesame seeds, which I shook on top. I took a bite. Chewed. Swallowed.

It was good. It was not rubbery, not fishy, not salty. It tasted fresh and healthful, much like seaweed salad—soy, sesame, and vinegar flavors mingling—but much less chewy, maybe even a little crunchy. The taste of the jellyfish was so subtle as to be almost nothing at all. I ate some more. It was a tasty, light, savory salad. In spite of all my anxiety about buying, soaking, preparing, and then eating it, jellyfish was completely unremarkable. I shook a few more sesame seeds on top and loaded another scoop of antioxidants and protein onto my fork.

4

Missing Polyp

Rather than sate my growing curiosity about jellyfish, our spring break to the Alabama coast and my foray into gelatinous cuisine only made me hungry for more. Something about seeking out jellyfish was giving me a new sense of purpose. Conveniently, I'd discovered great jellyfish stories throughout the San Francisco Bay Area. It wasn't hard to come up with reasons why traveling to the West Coast to learn more about jellyfish just made good sense: My sister offered to take care of Ben and Isy in exchange for our care of her three kids sometime in the future. Keith had some frequent-flier miles to use up. And we hadn't been away as just the two of us in a long time. I arranged to leave a couple of days early to do my reporting, and Keith would meet me afterward in the city for a weekend getaway.

The first meeting on my schedule was with Alex Andon, the founder and owner of Jellyfish Art, a supplier of jellyfish tanks and jellyfish. He suggested we meet at a hipster restaurant in a part of San Francisco called the Dogpatch. The neighborhood is equal parts residential and industrial; Jellyfish Art's offices were located in one of the semi-gritty warehouses that lined the waterfront across the street from the restaurant.

Alex received his undergraduate degree in biology from Duke and parlayed that into a job at a San Francisco biotech company. But when the tech bubble popped, he was laid off. During that lull, he visited Monterey Bay Aquarium's world-famous jellyfish exhibit, and upon seeing the crowds, he recognized a business opportunity. He made it his mission, as he explained, "to make jellyfish the next Pet Rock."

He began by simply putting up a website advertising custom jellyfish tanks called kreisels, which are rounded aquaria named for the German word for a spinning top. They are the best solution to the problem of building a suitable jellyfish habitat. Jellyfish bodies are delicate. Corners, bubbles, and filters mangle their tissues and tentacles. Strong currents, especially those that suck them toward outflow filters, are a major hazard. The kreisel current rotates in gentle circles, creating a watery cocoon for the jellyfish.

"The first year was a disaster," Alex admitted. He told me about jellyfish slipping behind filters, jellyfish clogging drains, about many, many leaky tanks, including one that overflowed, flooding his apartment and raining seawater on a downstairs landlord in the middle of a party. None of those challenges was a deal breaker for Alex. What drove him out of the custom jellyfish tank market was economics. With many competitors in the Bay Area, the margins were thin. And what Alex really wanted to do anyway was bring jellyfish to the masses. "So I started selling desktop tanks online. It really started to take off about a year ago." Alex designed his own desktop version of a kreisel. When we met, he was in the process of having it manufactured in China.

After lunch, Alex and I walked over to his warehouse. The facility wasn't huge, about the size of a basketball court, but things were humming. Plumbing for making seawater snaked around high shelves. A giant cylinder brimmed with a red liquid, colored by the maroon baby brine shrimp that were the main ingredient in Alex's jellyfish food. A string of homemade kreisels that lined one wall contained Jellyfish Art's stock of jellyfish. About a hundred moon jellies undulated in an elegant dance to the melody of running seawater.

Alex said that developing a breeding program for the jellyfish had proved more difficult than he'd anticipated. He was currently collecting animals from the ocean, but the permits were expensive. And once his new jellyfish tanks became available, he expected a much larger demand. He needed a bigger jellyfish supply.

Understanding what it takes to breed jellyfish requires a little jellyfish sex ed. So is jellyfish sex any good? Certainly there's no tedious dating. Jellyfish don't waste much (or any) energy on seduction and allure. But as far as interesting sex, jellyfish have that in spades. Here's how it works: A medusa, which is either a male or a female, produces either eggs or sperm. Some burp up their genetic material, cast it out into the wide-open ocean, and hope for the best. In other species, a mother jelly receives a packet of sperm from a male and broods her young among her tentacles. Either way, if a jellyfish sperm meets a jellyfish egg of the same species, new life begins.

But now the jellyfish life cycle veers away from the familiar.

The fertilized egg grows into a larva, called a planula, that one jellyfish scientist aptly described as looking like a furry Tic Tac. The fur is a rug of eyelash-like extensions called cilia. Once a newly minted larva has its cilia wiggling, it starts swimming. Upon bumping into a rock or a dock or a boat hull—or in the lab, a Petri dish—it performs a spiral dance that ends when it attaches itself to the surface. And then, like an enchanted creature, it begins shape-shifting. The larva elongates into a tube and grows wispy tentacles. It builds a digestive tract down the middle of the tube and a mouth in the middle of the tentacles. This being is now called a polyp. Unlike human infancy, the polyp is not a brief stage that the jellyfish passes through. A jellyfish polyp is a fully realized organism. The jellyfish can be a polyp for much longer than it is the more familiar medusa. The length of time varies for different species. Some jellyfish probably remain polyps for years or even decades.

Alex pointed out his breeding tank next to the kreisels, which held dozens of fuzzy Petri dishes sitting in still seawater. It was not the kind of fuzz Alex wanted. What he'd hoped for was a fuzz of moon jellyfish polyps. What he had was a coating of algae. Alex wanted moon jellyfish polyps to grow on those dishes as a way of getting the large quantity of medusae he needed to stock his tanks. "Once you have polyps, they make more of themselves," explained Alex. While the medusae reproduce sexually, the polyps reproduce asexually, and they do it very well. One polyp can multiply itself to make new polyps, sometimes fields of them, sort of like the way new shoots of grass proliferate in a yard.

But that's not all. Some polyps scoot along the seafloor and leave behind the most extraordinary footprints: A piece of the polyp's foot breaks off and remains where the animal had been standing. This dollop of polyp forms a tough coating around itself, at which point it is called a podocyst. When I first heard the word, I thought of it as a seedpod, but later I realized that the *pod-* comes from the Greek word for "foot." Still, a seedpod isn't a bad mental image. The podocyst can lie dormant for years until one day the conditions change, and it sprouts into a polyp, much like a plant sprouting from a seed.

As if shape-shifting, cloning, and forming foot seeds weren't enough, the next stage in the jellyfish's reproductive process is perhaps the most bizarre. At a cue from the environment, like changing water temperature or salinity, the polyp's stringy tentacles are absorbed into its body. The tube grows even longer, and what looks like a stack of pancakes forms where the tentacles used to be. The name for this process is strobilation. Forked branches extend from the center of each pancake, forming a delicate snowflake-like organism. The pancakes begin to wriggle individually. The forks flicker, and with what is sometimes described as a pop, the top pancake flutters away from the stack. Now the next pancake in the stack intensifies its quivering, its forked branches extend, and it too pulls away. This continues down the stack.

Thus freed from its sessile existence, the baby medusa is called an ephyra, a name shared with an ancient Greek sea nymph who lived in a silvery cave in the depths of the Aegean Sea. Ephyra came from hearty sea stock. She was one of three thousand daughters of Oceanus, the god of the world ocean, and Tethys, the sea goddess. So, too, this newly budded ephyra has abundant siblings. Depending on the species and the polyp's health, between one and sixty ephyrae bud from a single polyp. The budded ephyra pumps its short arms a few times and swims into the sea. It begins to eat tiny plankton and soon grows to about a third of an inch across. Its bell inflates. It grows tentacles. Oral arms extend from its mouth. If there's food available, the maturing ephyra is capable of growing an inch every few days. Soon the creature is recognizable as a medusa, and it's likely seen clustered with hundreds or even thousands or millions of creatures with very similar DNA: other medusae from the same polyp or its clones. Once strobilated, the polyp doesn't just roll over and die. Polyps have the potential to continue to clone and to strobilate over and over again. That's why Alex needed to grow polyps. Without polyps, he'd never get medusae, and without medusae, he wouldn't have a business.

After speaking with Alex, I left San Francisco and drove down the coast toward Monterey Bay, where I planned to visit the jellyfish exhibit at the Monterey Bay Aquarium the next day. I checked into a cheap hotel perched on the edge of Highway 101. Outside my window, the strawberry fields of California's agricultural industrial complex burst with color—deep green leaves and ripe red berries. I woke in the morning to unusual quiet. I was alone for the first time in a long time. There was no shower running, teeth being brushed, or toilet flushing. No sleepy husband. No grumpy kids. No pleading cats. Only strange silence. I flicked on the TV to generate some noise, and clicked through a few stations. I passed the *Today* show, paused on CNN, and then landed on the Discovery Channel, which announced the start of—I kid you not—a documentary called *Rise of the Jellyfish.* What were the chances? I settled in to watch.

Rise of the Jellyfish was astonishing. To really get an idea of the show, you have to say it in your head using your deepest voice and adding a bunch of reverb: *Riiissse of the Jellllyfiiiish.* The show portrayed jellyfish as demon creatures straight out of a B-movie. And it had overtones: overtones of a plot by a deranged Dr. Evil, or maybe climate change, amassing an army of jellyfish, creatures perfectly engineered to mortally wound us with their venomous toxin, to overrun our beaches, and to—oh yes, it's possible—destroy civilization as we know it.

What was really astonishing was hearing the deep-voiced narrator introduce one of the stars of *Rise of the Jellyfish*, none other than Dr. Monty Graham of Dauphin Island Sea Lab. I watched as he dissected a giant jellyfish, a species called *Nemopilema nomurai*, that lives off the coast of Japan. The animal was lying on the back of a boat. It was a dark reddish color and about three feet wide. Monty explained that despite its girth, the jellyfish had just two cell layers, as opposed to most other animals, which have three. Like the kids on the beach in Alabama, he sliced open the jellyfish and pointed out the jellylike internal material, the mesoglea.

At this point in the documentary, the narrator could have said a lot of very interesting things about mesoglea. Mesoglea—the very spinelessness of the jellyfish—is a big part of what gives them so much of an advantage, not just in today's seas, but also for more than half a billion years.

The narrator could have pointed out that mesoglea allows jellyfish to move through the water efficiently. Imagine a water balloon. When you squeeze it in the middle, the water bulges out toward the ends. That's because water, unlike cotton or marshmallows, is incompressible. It always takes up the same amount of space. When you release the middle of the balloon, the skin of the balloon pushes on the water, and it naturally returns to its original shape. A jellyfish is much like a water balloon. Its muscles squeeze against the watery mesoglea contracting the bell. When those same muscles relax, the incompressible mesoglea surrounded by a thin skin returns to its original shape. The watery mesoglea provides this service basically for free. Hardly any muscle energy is required to return the bell to its relaxed position.

The narrator could have also pointed out that mesoglea gives jellyfish a deal on buoyancy. Flotation is really important if you live a couple of miles above the nearest solid surface that could catch you if you fall. Animals are made of proteins and carbohydrates, which are heavier than water. An animal with a hard skeleton, like a shrimp or a fish, has to compensate for its load, which can be three times denser than water. Animals either have to swim constantly to stay afloat, which takes a lot of energy, or they can come up with a scheme to create a float using some part of their bodies. Bony fish dispense some of the gas they breathe into a swim bladder that acts like an inner tube. Other animals fill parts of their bodies with fats, which float in water. Ingeniously, jellyfish convert their mesoglea into a float by actively controlling its chemistry. They pump out heavier ions in exchange for lighter ones with the same electrical charges. In moon jellies, heavier sulfate ions are traded out for lighter chloride ions. There's so much trading done that the mesoglea contains only half as much of the heavy sulfate ion as the surrounding seawater. This ion trade offsets the heaviness of the proteins and carbohydrates to the point that the density of the moon jelly is nearly the same as the density of seawater, making floating easier and falling into the abyss much less likely.

The narrator did note that jellyfish can tolerate polluted zones in the ocean. Studies of sea nettles in Chesapeake Bay showed that jellyfish were insensitive to oxygen concentration, growing just as well in water with low oxygen as in water with normal amounts of oxygen, but I didn't come away from the show understanding that it's the remarkable chemically controlled mesoglea that gives them that ability. Just as sailors in submarines control the gases inside so they can survive in an unnatural environment, so do jellyfish, storing in their mesoglea the dissolved gases they need to survive.

No, the baritone narrator of *Rise of the Jellyfish* didn't extol the extraordinary capabilities of the mesoglea. Instead his message was that even with just two cell layers, jellyfish are like a deranged army rising from the deep in order to sink our ships and sting humankind into submission.

Monty's unfriendly reception made some sense now. There was no nuance in this documentary. No wonder Monty asked if I had an agenda. The show sensationalized jellyfish and used fearmongering for attention. It used Monty's name and expertise to justify its message, and I began to understand why he might be cautious about my intentions.

The Monterey Bay Aquarium was one of the first public aquaria to exhibit jellyfish, and until recently, one of the few in the United States to have permanent displays. The original genius behind the jellyfish exhibits was a Harley-riding martial arts expert, Chad Widmer, a man with a translucent thumb. Chad pioneered methods to coax the jellies through their complex life cycle in the lab and developed much of the equipment to make it happen. He shared his knowledge with others, in part by writing *How to Keep Jellyfish in Aquariums*, which is such a quick and good read that I devoured this handbook on jellyfish husbandry cover to cover. Chad has called his philosophy of jellyfish-rearing *kuragedo*, Japanese for "the way of the jellyfish." He summarized it in a column written for an international conference on jellyfish husbandry:

> A white belt in Kuragedo can walk around the gallery without accidentally destroying anything. A yellow belt can harvest *Artemia nauplii* and rotifers, and feed them to the appropriate animals. A green belt knows the scientific names of things, and can bleach a kreisel whilst a black belt can start new cultures and sustainably keep a gallery going. A master dreams up new things that make us all reach for our cameras. As for the color of my belt . . . all of the black has worn off, leaving me a white belt once again.

Chad had recently left the Aquarium to work on a PhD on jellyfish and climate change in Scotland, but I had arranged to meet with Chad's successor, Wyatt Patry. I arrived an hour early at the remodeled cannery that now houses the Aquarium, so I'd have time to browse the exhibit. Unfortunately, I had arrived between big exhibits and had to settle for a single tank full of jellies tucked into the side of an exhibit on climate change. In a cylindrical tank, illuminated beautifully, jellyfish pulsed upward and downward, sometimes lazily tumbling over one another. The top of each brown bell was speckled with a number of perfectly circular white spots, giving them the appearance of enchanted toadstools. This species was the spotted jelly (*Mastigias papua*), which hosts symbiotic algae in its skin like a photosynthetic tattoo.

I sat down on a nearby bench to observe. Most of the people visiting had children. Invariably, the kids spotted the tank and ran toward it, squealing, "Jellyfish!" If their first language was Spanish, the cry was, "Medusa!" The shouts contained glee, some surprise, and a bit of longing. The kids almost uniformly stretched their hands toward the glass as if to touch the jellies. Another common cry was "Baby!" with a finger pointed toward the tiniest of the jellyfish. But of course, these weren't the real babies at all. The real baby jellyfish, ephyra, were even smaller, and even cuter, than these kids knew. I was about to get to see them.

Wyatt Patry is tall and clean-cut. He has the coolness of a surfer and the calm reliability of a lifeguard. He told me that when he was a kid, his parents brought him to the Aquarium all the time. "My parents loved the jellyfish exhibit. But I just kind of ignored it and breezed through the gallery. I liked the sharks. So I don't know when I started to like jellies. I maybe took longer to appreciate the, you know, grace. When I was a teenager, I wasn't that into the whole graceful thing." But now he's hooked, and it's partly because he enjoys the challenge of raising an animal with such a complicated life cycle.

Patry invited me backstage at the aquarium, where they were prepping for the new exhibit, which was still closed to the public. A dozen purple-striped sea nettles trailing oral arms like wedding veils turned lazily in a huge kreisel. Wyatt said that until recently it was the largest kreisel in the world. To access the tank, you had to climb a ladder that extended up two stories. But from the side, the tank was just a foot thick. Spotlights focused through the narrow sides created the artistic display from the front. Mood music filled the room. It felt more like an art gallery than an aquarium.

We headed farther back into the aquarium, through a locked door and into the labs where the jellyfish were bred and raised. Gone was the soft lighting and the atmospheric music. The floor here was linoleum, the lighting fluorescent, and the sound the roar of motors, filters, and rushing water. On one side of the lab, the tanks were smaller and the water flow was slower. This was where breeding happened. I noticed that in contrast to Alex's breeding setup, Wyatt's Petri dishes were upside down and propped up on one edge. I made a mental note to tell Alex to prop up his dishes. "Why do you do that?" I asked.

"Because the polyps like to grow on the undersides of things. Look here." Wyatt pointed to a tank full of the same type of translucent jellyfish I'd seen in Alex's facility. "These moon jellies were collected yesterday. See the purple fluffy stuff? Those are planulae." Moon jellies brood their larvae in their tentacles. Lavender clouds floated under some of the animals' translucent bells.

Wyatt showed me overturned Petri dishes in a nearby tank. "Those are moon jelly polyps. They're all popping off right now."

I peered down through the water in the tank and saw furry polyps coating the undersides of the plastic dishes. As my eyes focused down to the realm of the barely visible, I suddenly recognized a snowstorm of ephyrae. Hundreds—no, thousands—of filigreed bits of life pulsed upward in the tank. As each was born, it did seem to shudder to life with a pop. They were so delicate, and yet so assured in their pumping.

"Oh my god! They are so cute! Look at them. They can swim. Oh my god! They are so cute!" I sounded a lot like the kids I'd watched squeal at the jellyfish exhibit in the public halls downstairs.

But I forgave myself for my enthusiasm. They were swimming heartbeats. I was watching magic.

I first found Lucas Brotz in my daily Google alert. Lucas was a PhD student at the University of British Columbia and had published a controversial paper on the growth of jellyfish populations in coastal regions around the world. Some scientists heralded the paper as a first step toward understanding jellyfish on a global level. Others criticized Lucas for using anecdotal information from fishermen and other nonscientist observers. The media were predicting the demise of the ocean on the basis of his study. I scheduled a conversation with Lucas to find out more, and we discussed the jigsaw pieces of climate change, overfishing, pollution, and invasive species that make up the jellyfish puzzle.

Lucas was working on a paper about jellyfish polyps at the time. He said, "We don't know that much about the polyps, but without a polyp, you don't get a medusa. Polyps are really just little jellyfish factories. My personal opinion is that one of the factors that will turn out to have a big influence on jellyfish populations is the increase in hard surfaces for the jellyfish polyps to live on from our construction of artificial structures on the shoreline."

Coastal development is rampant. Worldwide, we've built more than 14,000 commercial harbors. We've dredged 4,100 kilometers [a little over 2,500 miles] of waterways. We've constructed over 8,300 oil rigs offshore. Aquaculture farms are increasing production by 7.5 percent per year. The number of wind energy structures has been increasing by almost a third each year. Thirty-four countries have built artificial reefs. The numbers of buoys, moorings, breakwaters, and stretches of concrete shoreline are innumerable. And all of those hard surfaces, with untold acres of undersides, are potentially brand-new habitats for jellyfish polyps.

Lucas mentioned one of Jenny Purcell's studies of wild moon jelly polyps in Puget Sound, which I looked up later. At a small recreational marina, she discovered a meadow of moon jelly polyps on the undersides of the boat slips. Using scuba, she and her collaborators photographed and counted the individuals at several different spots around the marina. In every square centimeter, an area the size of a penny, they counted about nine polyps. For three years, the scientists dived under the docks, monitoring polyps. Each polyp split into an average of ten pancake-like discs. Between 40 and 86 percent of the polyps strobilated at any given time. The dock had a footprint about the size of a three-bedroom house and an estimated population of 100 million polyps. A back-of-the-envelope calculation said that just that one dock produced over 500 million medusae, more than the human population of the United States!

Jellyfish scientists have compared the jellyfish polyp to a Trojan horse. By building habitats for the polyps, we are inviting them into our bays and ports and beaches. They arrive and plant themselves, appearing as benign guests. And then, when they get the cue, they release an invasion of medusae before we can react.

There's one huge caveat, however. Despite all the artificial jellyfish habitats we've constructed, we've observed the polyps of only a minuscule number of jellyfish species in the wild. Of the thousands of species of jellyfish, what we know about the vast majority of polyps comes from experiments in the lab or from bays near marine stations. Published observations of polyps in the wild cover just over two dozen species. Moon jellies, the most cosmopolitan of jellies, are the exception. They are frequently seen coating the bottoms of docks, like the one in Puget Sound. Millions of giant jellyfish swarm the coast of Japan, but only a handful of polyps have ever been seen. Entire fisheries profit from two species of edible jellyfish, but the locations of the polyps that produce those medusae remain mysterious. We don't have accounts of wild polyps of the box jellies that brutally sting swimmers in Australia every summer, of the spotted jellies that can proliferate into swarms of millions of medusae in the Gulf of Mexico, of the stocky barrel jellyfish that surf the seas of the United Kingdom, or of the invasive nomadic jellyfish that blooms by the kilometer in the eastern Mediterranean.

Part of the problem, the scientists have told me, is that polyps are small and hard to see among the algae, anemones, bryozoans, barnacles, and other encrusting organisms that grow jungle-like on underwater structures.

Scientists have learned that when vision fails, DNA can sometimes give us answers, but jellyfish polyps remain elusive even when we throw biotechnology at them. Scientists in the Netherlands recently used DNA markers to try to suss out jellyfish polyps in the North Sea. The region is home to at least five types of moon jelly medusae, along with stinging nettles, lion's-mane jellies, and barrel jellyfish. They sampled twenty-nine locations near marinas and on more open coastlines. They looked at docks, oil rigs, and boat wrecks, and they even moored plastic plates for polyps to grow on. Among their 183 polyp samples, the only jellyfish DNA they found belonged to one species, the moon jellyfish. The researchers wrote, "The whereabouts of polyps of the other species in the southern North Sea region thus remains an open question."

We've looked, probably not as hard as we could, but we have looked. We just don't know where all the polyps are. They must exist, because we see lots of medusae of many different species, but where the polyps are remains one of the biggest mysteries in jellyfish science. Moreover, the mysteries surrounding wild polyps won't end once we find them. Lots of questions remain unanswered: What do wild polyps eat? What eats wild polyps? How long do they live? Why do they die? What causes them to clone, to produce podocysts, to produce medusae?

But for now, whenever I look out into the sea, I'm no longer immediately taken by the sparkle of the water or the power of the waves. The first thing that I think is, *Where are all the polyps?* We just don't know. And that's eerie.

In Jelly Genes

When it comes to jellyfish, it's not just the polyps that are hard to pin down; it's the medusae too. Jellyfish are among the oldest animals in the world, but for the amount of time they've been here on Earth, they haven't left much to show for it. There are vanishingly few jellyfish fossils. At the Denver Museum of Nature & Science, paleontologist James "Whitey" Hagadorn told me, "Finding *T. rex* bones is like shooting a fish in a bucket, because they are already mineralized. Finding a bag of goo that has no hard tissues, even partially mineralized, is like finding a needle in a haystack. Jellyfish fossils are the ultimate fossils in the sense that they are the rarest and the least likely to be preserved."

After my trip to the Bay Area, our family took a summer trip to Colorado to avoid the blazing summer heat of Texas. I took the opportunity to peel off from Keith and the kids—they assured me they were more interested in taking the cable car up Pikes Peak than in learning about jellyfish at a museum—so that I could catch up with Whitey to talk about ancient jellyfish.

Whitey and I had been friends in grad school. Students in my biology department and his geology department sometimes had joint happy hours on the roof of our building, drinking beer out of plastic cups among the air conditioner units and ventilation chimneys. He is tall and lanky and kinetic and, as his nickname implies, pale, with a long, white ponytail and a white goatee. I've never been in the field with Whitey, but it's always been easy for me to picture him striding across rocky outcroppings in a wide-brimmed hat, eyes focused downward, on the lookout for fossilized evidence of ancient jellyfish.

The oldest verifiable medusae fossils are about 510 to 500 million years old. Most of these appear to be typical jellyfish with a round bell, tentacles, and oral arms. There was a hot spot in jellyfish fossil preservation just over 300 million years ago, and then another one while the dinosaurs were clambering over the land 199 to 145 million years ago. In these more recent jellyfish fossils from the time of the dinosaurs, you can see a little more detail: gonads, tentacles, even muscles. One looks strikingly similar to the moon jelly, perhaps the most ubiquitous jellyfish on Earth today. But outside those sporadic time periods, few jellyfish fossils are to be found.

Despite the lack of abundance, the jellyfish fossil record does tell one powerful story about the ancient world. If jellyfish weren't the first creatures to pull themselves up off the seafloor, harness the buoyancy of water, and use muscles to swim, they were pretty close. In 2015, researchers from the University of Cambridge discovered a 560-million-year-old fossil in a cave in Newfoundland, Canada. Unlike any other fossil of that age, parallel rows ran through the stone like rake lines in the sand. They were the oldest muscles ever found, predating the previous record holder by 100 million years. The discoverers believed the primitive muscles belonged to a type of jellyfish.

The scientists named the creature *Haootia quadriformis.* In the language of the indigenous people of Newfoundland, *haoot* means "demon," and the fossil—with wavy forked tentacles—has a sort of sinister appearance. *Quadriformis* denotes its four-way symmetry, a feature shared with today's jellies and corals. What was amazing about the fossil was that its muscles looked a lot like those we see today in a parasitic jellyfish called *Polypodium hydriforme* that lives inside the fish eggs we call caviar. But its overall shape was similar to a small group of jellies called staurozoa, which look like stemmed wineglasses with tentacles around the top edge. Today about fifty species of staurozoa live in our seas. For decades everyone thought they were small in number and favored cold waters close to the coast. But in 1998, during deep-sea-submersible dives to hydrothermal vents in the Pacific Ocean, scientists discovered entire ecosystems where fields of staurozoa flourish.

No one knows exactly how ancient jellyfish first rose from the seafloor and began to swim. I like to imagine that the first swimming jelly looked something like *Haootia*, that a strong current pulled the muscular top from its stem, the way dandelion florets are pulled off by the wind. Tentacles pulsing and muscles contracting, it managed to wander quite a distance from its parent. Perhaps when it settled somewhere new, it found a rich, unexploited habitat, and it grew strong there, producing more stalked jellies with the same type of muscular tops that could be pulled off by currents. These proto-medusae inherited their parents' ability to float-swim to new, fertile lands. Evolution favors this type of behavior. It's the way new innovations, things as basic as movement, become a fixture in the world.

"What's the one burning question you have about ancient jellyfish?" I asked Whitey after we'd been talking a while.

"What they ate," he answered without hesitation. Right. What *did* they eat? The jellyfish we know today mostly eat shrimplike plankton or fish eggs, which didn't evolve until 50 million years or so after the most ancient jellyfish. No one has found any food in the stomachs of fossil jellyfish.

Maybe they ate nothing. Maybe jellyfish came to be in a time when creatures absorbed what they needed from the world around them and then recycled it. Maybe the time of the jellyfish's origin was an era before hunting, before aggression, before violence. Maybe when we think of prehistoric times, we should recall not a fierce place where dinosaurs stalked the Earth, but a time before that, a time of peace.

About half a billion years ago, creatures started burrowing down into the sand, disrupting microbial mats. Mouths hardened, jaws formed, teeth developed. Hunting began. Skeletons formed in response to predation. Spikes grew from the newly hardened shells. Claws formed to overcome the toughening of outer body layers. A climate change occurred, one that increased oxygen levels, decreased carbon dioxide levels, and like today's changing atmospheric chemistry, also changed the chemistry of the sea. All these things, and perhaps others that we haven't yet teased from the ancient geology, changed the world. Most of the soft, mushy creatures that were alive when jellyfish first appeared couldn't survive this hostile new environment. A mass extinction occurred about 555 million years ago, killing off all but a few of the ancient creatures. But the comb jellies and the jellyfish survived.

So maybe, when we look at jellyfish, what we see is a reminder of our deepest origins. Perhaps jellyfish strike an unconscious nerve, far below what we are still certain we know, of a past before violence, before consumption, before aggression. Perhaps jellyfish are living ghosts of a kinder past, a ghost from our true garden of Eden.

Not unlike the problems paleontologists have with jellyfish, lack of evidence also plagues scientists working on modern jellyfish. A lack of hard parts means that the anatomy of jellyfish can be, well, mushy. During my trip to Dauphin Island Sea Lab, I had sat down with Keith Bayha, a jellyfish geneticist. Keith gestured to a creature preserved in a jar on his desk. "That's a *Pelagia*. *Pelagia* has a pretty common body plan. You can tell them apart from a sea nettle, but when it comes to really defining things, it gets a bit wonky." He explained that when you have two similar-looking jellyfish, it's not always easy to decide if they are the same species or not. Maybe they have nearly identical body structure, the same number of tentacles, the same shape of stomach and digestive system, the same placement of sensory organs. But one's tinted blue and the other's flushed pink. Are they different species, or were they just eating different types of food? Or are they the same species, but one has hit middle age, while the other is prepubescent? That's where you can turn to DNA. The DNA of an individual doesn't change with age or environmental conditions. So scientists like Keith are starting to use DNA to solve mysteries about jellyfish that have long been cold cases.

Keith pointed to a photograph of a large, frilly jellyfish on the wall behind me, one that is sometimes called the pink meany. Until Keith cracked it, the identity of this jellyfish had been a cold case for well over a century. And the mistake had started with one very famous biologist.

In his later years, Ernst Haeckel became one of the most influential and recognized scientists of all time, as well-known as Darwin during their contemporaneous lifetimes. He named and drew thousands of new species, mapped a tree of life that related all known species, and established the kingdom that contains single-celled creatures, the Protista. His stunning and ornate drawings of plants and animals—including the gorgeous jellyfish illustrated on the jacket of this book—helped launch the Art Nouveau movement, and posters of the illustrations in his books are still in print. In the 1870s in Jena, Germany, Ernst had already begun to gain celebrity as both a zoologist and an artist. One day he sat in his laboratory dissecting a certain frilly pale pink jellyfish that had been collected off the Dalmatian coast of Croatia during a research trip in 1871. There's no record of that dissection, but we can imagine it.

Wooden shelves stocked with colored-glass bottles of chemicals line the walls. Sunlight streams through thick windows illuminating delicate forms suspended in clear glass jars. Haeckel locates a particular jar containing a certain jellyfish, and as he unscrews its lid, the acrid smell of formalin wafts through the air. He carefully removes the jellyfish and lays it bell side down on a dissection tray, exposing its underside. Pendulous oral arms fall back upon themselves like lavish curtains. These frilled, tube-shaped tissues are extensions of the jellyfish's mouth, like long lips, and so they are called oral arms. This animal has eight. There's only one way in and one way out of a jelly's body. Complete guts, those that process food as it passes in one hole and out the other, evolved after jellyfish did. In a jellyfish, food enters the mouth, gets digested in the stomach, and then passes around the body via a series of branching canals, supplying energy to the strong muscles that encircle the bell. The unused waste follows the reverse path and is ejected out the mouth.

Selecting a sharpened scalpel from his tool kit, Haeckel skillfully dissects away the oral arms, exposing thin tentacles that tangle together like cooked spaghetti noodles. The tentacles are densely packed with stinging cells, the animal's only weapon, used to capture food. Haeckel makes a few more strokes in a sketchbook. With the tentacles and oral arms removed, he has a clear view of eight small sensory organs called rhopalia, which jellyfish use to see light and darkness, to feel currents, and to sense gravity. They dot the underside of the bell like pinheads poking out of a pincushion. Again, Haeckel draws. Then he carves away a layer of tissue. The slice exposes the clover-shaped band that contains the jelly's gametes. Medusae are either female or male and so these organs contain only either eggs or sperm cells.

Now I imagine Haeckel pushing his chair away from the table, stretching his back, which has cramped from leaning over for so long. He wriggles his fingers to loosen them after clutching the scalpel and the pen. Haeckel rises from his table, carefully washes and dries his hands, and returns to his notebook. This jellyfish is a species new to science, so it's Haeckel's job to slip it into the known tree of life, but naming this animal gives him some trouble. He writes, "I had this strange, in many respects very strikingly different Medusa. . . . This new genus differs so much from the remaining Cyaneidae [the family it most closely resembles] that I propose a special new subfamily (or even family?)." However, for some unknown reason, Haeckel reconsiders. Instead of creating a new family, he places this jellyfish in the existing family of Cyaneidae—a misidentification that remained with the animal for more than a century.

In his windowless office at Dauphin Island Sea Lab, Keith Bayha continued the story: "Haeckel was great and all. And it's great that college kids put his posters on their walls, but if you look at his science, he made a lot of mistakes. But Gustav Stiasny—that guy was always right."

In the early 1900s, Stiasny was a biology student at the University of Jena, Germany, where Haeckel had by this time become somewhat of a celebrity. He was one of the era's most popular orators, often mixing politics with natural science, generating rich material for students to argue about in the bars around the university. In 1903, Stiasny left Jena and returned to his hometown of Vienna, where he earned a PhD. Taking a cue from Charles Darwin's adventures aboard the *Beagle*, for the next decade Stiasny sailed on scientific expeditions to Greenland, the Azores, and the Canary Islands, as well as across the Atlantic to Argentina, Uruguay, and Brazil. Stiasny landed at the Zoological Station in Naples in 1913 and was appointed head of the department of plankton and fisheries research. He soon met a Dutch biologist named Gerarda Wijnhoff, who was at the station studying nemerteans, a family of oceanic worms whose Greek moniker is derived from the name of the sea nymph Nemertes, considered the wisest of the fifty Nereids. Wijnhoff and Stiasny married two years later and moved to the Netherlands, where Stiasny became a curator at the Rijksmuseum of Natural History in Leiden. Stiasny and Wijnhoff published prolifically, sometimes together, sometimes separately, on an astounding array of marine animals. Stiasny authored or coauthored papers that named more than a hundred new species. A description of Stiasny characterizes him as a man with "a bubbly personality" who had "huge and limitless enthusiasm" for his work. He was musical, sometimes accompanying his lectures with songs to make a salient point. The students loved it. Though his lectures weren't required, "they were still popular because of their originality."

And then, everything changed. "In 1940, when the German heel crushed our country and anti-Semitic laws and measures were forced upon our population," wrote Wim Vervoort, Stiasny's colleague at the museum and his obituary writer, Stiasny "was compelled to wear the notorious yellow star and had to leave the Rijksmuseum." It turns out that Stiasny was Jewish during a time when to be defined as Jewish was to be persecuted. Despite his brilliance and his commitment to science, he was deemed unfit to keep his job, to participate in society, to live. What could that have been like for Stiasny? He was a man devoted to identifying the unique characteristics that define an animal, only to learn that a single characteristic—one that had nothing to do with his ability—would be his ruin. A single characteristic defined him as less than a human.

During the first year of the war, Stiasny was forced to wear the yellow star and leave his position at the museum. Still, the prolific biologist churned out eleven scientific papers. Two of these dealt with an animal he'd hoped to study for quite some time, that curious pink jellyfish from the Dalmatian coast. He wrote, "For many years I've been on the hunt for *Drymonema dalmatina*, one of the most beautiful but unfortunately also the rarest scyphomedusae in the Adriatic and the Mediterranean. . . . At last I have succeeded . . . to come into the possession of one specimen." It's thought that Stiasny managed to get ahold of a jar containing *Drymonema* from the Hungarian National Museum. He expanded on Haeckel's two-sentence definition of the species in fifteen pages of detail. He corrected Haeckel where the description was inaccurate, vague, or simply missing. Toward the middle of his exposition, Stiasny wrote, "As you can see, I'm in disagreement with Haeckel on almost every point."

Consider the courage required for Stiasny to publish those words. He wasn't just challenging the work of a respected scientist and former teacher. Given the political situation, the words were probably an affront to someone on the other side of the racial laws. Haeckel had been Aryan, correct, by definition. To write such words, Stiasny must have had a certitude that his science was more true than the regime that was rising to power around him. Perhaps the words even reveal the defiance of a man under attack for the religion that defined him.

Stiasny hid inside the Rijksmuseum during most of World War II. Fellow researchers risked their own lives by bringing him food. Mentally, he seems to have created his own sustenance, writing about two dozen papers while in the bowels of the museum. There's a suggestion that Stiasny's wife, who came from a non-Jewish and aristocratic family, played a role in his avoiding detection. But the war broke Stiasny anyway. He escaped the worst of the Nazi atrocities, only to die just one year after liberation. A scientist later evoked the atmosphere at the Rijksmuseum under the Nazi regime: "Let us recall that not just working facilities in the Museum, but food supply and personal safety, were severely compromised. . . . Quite probably these appalling vicissitudes contributed to the death in 1946, for example, of poor Gustav Stiasny. The story of what went on during the war years . . . is largely untold."

The episode has a poignant coda. The single specimen of the curious jellyfish that Stiasny obtained and then so carefully studied appears to have been a war casualty too. It seems that the animal was dutifully returned to the Hungarian National Museum, then lost in the bombing of Budapest during the uprising against the Communists in 1956.

"The occurrence of *Drymonema dalmatina* is characterized by periodicity," Stiasny wrote while examining the jellyfish in 1940. "The medusa was found on the Dalmatian coast in 1879, and then again in 1908, finally in 1937 at Rovinj, i.e., in an interval of about 30 years." On schedule, a little more than thirty years after its 1937 sighting, *Drymonema dalmatina* returned again, this time on the other side of the Atlantic, near Puerto Rico.

The marine field station of the University of Puerto Rico is located on the small island of Magueyes, just off its southwestern coast. In December of 1974, Ron Larson, a master's student at the field station, was told that a rare medusa might be present in the channels between barrier islands. Puttering through a small bay in a skiff, Larson noticed one rosy blob, then another, and another. Soon he counted a dozen, ranging in diameter from that of a frying pan to a trash-can lid. He pulled on his fins and mask, put his snorkel into his mouth, and flipped into the balmy sea. A minute later, Larson was face-to-bell with a jellyfish a foot and half in diameter. A mass of ruffled oral arms bulged below the bell like petticoats from beneath Little Bo Peep's skirt. These feeding structures extended well over a yard downward. Tentacles the thickness of a coffee stirrer stretched many feet in all directions. The animal hung in the water just a few feet below the surface. It pulsed very slowly, only two or three times a minute. Dozens of tiny fish schooled in a circle around the jelly's oral arms like sleek silver mounts on a merry-go-round. As Larson neared, the fish swarmed closer and closer to the medusa, until they were swimming among the animal's oral arms, apparently unharmed.

Larson noticed something white and gelatinous within the mass of curtainlike tissue that made up the creature's oral arms. He realized it was another jellyfish, a moon jelly of the genus *Aurelia*. The larger jellyfish had captured it, immobilized it, and was digesting it with its oral arms, not even bothering to pull it inside its mouth. This medusibalism is not unusual. Jellyfish are commonly thought of as eating zooplankton and fish, but many gain much of their nutrition by feasting on other jellyfish. Larson collected the jellyfish and took it back to the lab, taking the first photo of this mysterious animal. And then, *Drymonema* disappeared again. No one saw it for thirty years.

In 2003, Keith Bayha was on vacation in the resort town of Foça, Turkey, a picturesque fishing village nestled deep in a bay on the Aegean Sea. The town takes its name from the ancient Greek word for "seal," and endangered Mediterranean monk seals can still be seen there, occasionally popping their bowling-ball heads above the turquoise waters. The town is said to be the home of the rock on which Homer's sirens basked, which makes you wonder what role those seals might have played in the mythology.

At night, wooden fishing boats are tied up at the waterside of a stone boardwalk. Bright blue tarps are secured over gear; ropes and woven baskets litter the decks. Restaurants line the other side of the walkway, colorful awnings and umbrellas covering tightly packed tables and chairs. It creates a cozy space for dining on tender calamari and fresh sea bass, along with a crisp local wine. Keith had finished just such a meal. He strolled along the water's edge, taking in the town's charm. As jellyfish scientists do, Keith peered down into the water. His eye was drawn to a few pale pink shapes of jellyfish bobbing beneath the surface. But something about these jellyfish seemed different from anything he'd seen before. He needed to collect them.

Jellyfish must be preserved quickly in order to maintain their integrity for genetic analysis, but Keith was nowhere near a lab stocked with chemicals. MacGyver-like, he found a nearby pharmacy and bought several bottles of lemon-scented hand sanitizer. He scooped a few of the gelatinous creatures from the balmy water and submerged his specimens in the makeshift preservative.

After grad school, Keith joined a research group led by Michael Dawson at the University of California, Merced. Mike Dawson and his colleagues have helped untangle what is often a confusing family tree of jellyfish relationships using genetics. For this analysis, fifty-five different jellyfish from all over the world, from Alaska to Australia, from Slovenia to the Sandwich Islands, and from Norway to New Guinea, were used. Mike's group extracted and sequenced the DNA from each jellyfish. Then they fed the sequence data into computer programs that count the differences between one sequence and another.

Around the same time that Keith collected the curious jellyfish in Turkey, the same species appeared in the Gulf of Mexico, not just ruining its pattern of showing up every thirty years but also blooming to population sizes in the millions. Keith and Mike were able to look more closely not just at the DNA, but also at the jellyfish's anatomy. They found that the animal was unlike all other jellyfish, both in its DNA and its appearance. The genetic and anatomical features of the jellyfish were so different that it didn't qualify for inclusion in any of the known jellyfish families. If Keith and Mike could have put the Hogwarts sorting hat on the creature's bell, it would have said that it was time for the school to expand and introduce a new house of wizards. When the analysis was published in 2010, the jellyfish became the definition for a new family, called Drymonematidae, the first new family added to the Scyphozoa—the class that includes the moon jelly and the sea nettle, the animals you think of as typical jellyfish—since 1910. It was also the first new family added to the entire jellyfish clan—which includes the box jellies, the mostly small clear hydrozoan jellies, and the stalked staurozoans—in ninety years.

Today Ron Larson, who studied the ecology of the pink jellyfish of Puerto Rico in the 1970s, works for the U.S. Fish & Wildlife Service in Oregon. He still focuses on rare creatures, though unfortunately they are rare because humans have made them so. Larson helps find solutions to questions of how to best protect endangered species in Oregon rivers. One afternoon, his office phone rang. Keith Bayha was on the other end. He introduced himself, then asked, "Do you remember that jellyfish you were working on in the seventies?"

Of course Larson did. Keith described his work on the jellyfish. He told Larson that the genetic and morphological differences between the frilly pale pink jellyfish and all the other jellyfish were vast: "Drymonema is in a whole new family," he said. The rejiggering of the jellyfish taxonomy meant not just a new family, but a new scientific name for the species as well. "And the species name is *Drymonema larsoni*." The animal finally had its definition.

The considerable differences between *Drymonema larsoni* and its relatives, as well as the fact that those differences had slipped from notice for more than a century, illustrate how hard it is to study jellyfish and as a consequence how little we understand about jellyfish species.

Keith pointed to a picture of a moon jelly, one of the most common of all jellyfish and probably the one with the longest pedigree of scientific papers. He told me that this animal was a member of a new species that he and another biologist, Luciano Chiaverano, had recently discovered in the Gulf of Mexico through genetic analysis. "There are two moon jelly species living side by side. One's kind of pinkish. One's kind of grayish. But they are not different enough that you'd say, 'Oh that's definitely another species.' There were three to five recognized species of *Aurelia* before we started looking at their genetics. But there are really like twelve, thirteen, fourteen species—a lot more."

A species that looks a lot like another species is called a cryptic species, and scientists have now teased apart a handful of cryptic species among the animals that look like moon jellies. They are so new that they haven't been given species names but are just called species 1, species 3, species 4, and species 6. *Aurelia* species 1 is the moon jelly that lives in the Pacific Ocean from Australia through China, Korea, Japan, and into California. This population has been on the rise since the 1980s near Japan and since the 1990s in Korea and China. It's the culprit in the very first jellyfish-induced power outage, which blanketed Tokyo in darkness in the 1960s. Species 3, 4, and 6 are also found only in the Pacific. What was thought to be one moon jelly species in the Mediterranean turns out to be three. Defining the moon jellies is very much a work in progress.

The best bet is that the number of jellyfish species is probably double the official list. We've named about 300 species of true jellyfish, but the total number is probably more like 600. About 50 species of box jellies and another 50 siphonophores—colonial jellyfish like the Portuguese man o' war—are described, but there are probably 100 of each. And 2,000 species of hydrozoans—the mostly tiny, mostly clear jellyfish that are the most numerous of all—have been officially recognized, but there are likely 4,000. The diversity among jellyfish is very much underestimated.

How all that diversity plays out in the ocean is also poorly understood. As Keith Bayha and Michael Dawson wrote, "Cryptic taxa . . . likely result in cryptic ecology." Since different species respond to environmental changes differently, species diversity means that there are more factors to juggle when trying to predict what a future ocean might mean for jellyfish. Things like temperature, salinity, acidity, pollution, and overfishing can affect different species in vastly different ways. High temperatures and low-acidity water might cause one species to explode in numbers, while those same conditions may cause another to die off. A third species might be impervious to changes in the temperature or the acidity. For a scientist trying to understand whether or not jellyfish numbers are on the rise globally, this is a real predicament. How can we predict what jellyfish will do in a future ocean, if we don't even know what's out there?

And here the story circles back to the ancient world of jellyfish, because DNA isn't useful only for telling us about what's in the ocean now. It can teach us about what was in the ocean hundreds of millions of years ago too. DNA has been with us as long as life has been life, so it's a kind of molecular fossil. The various nucleotides in the chain of DNA are thought to mutate at more or less steady rates. The more differences there are between two different species, the more time it took for those differences to accumulate. Scientists can build a history based on how different species are from one another, then use fossils (if there are any) and anatomy to pin down the specifics.

For hundreds of years, sponges were thought to be the animals with the most ancient origins. That's because they have simple bodies, even simpler than jellyfish or comb jellies, which have actual muscle bands and nervous systems. A dominant idea in evolution was that animals develop from simple to complex. Once complex cells and tissues like nervous systems evolved, they wouldn't disappear because having a more sophisticated tool gives an animal a survival advantage. (Parasites are an exception. They often give up their complex systems and instead rely on those of their host to navigate the world.)

But in 2008, scientists delving into the origins of animals performed an analysis that called the ever more complex theory into question. Comparing the DNA of a few different genes across a large range of animals suggested that it wasn't sponges at the base of the animal tree of life. It was comb jellies. Remember the comb jelly *Mnemiopsis*? The same animals that took over the Black Sea seemed to have more differences in their genetic code than any other group of animals. This suggests that they were the first to have split from the evolutionary tree and that, among all the animals still alive today, they are the oldest. The scientists published the result, but cautiously. It was just one analysis, after all. But then in 2013, another group repeated the work, this time using the entire genome—all the DNA—of the comb jelly. The results agreed with the first study.

The implications were significant for how we view life, and brought up questions that were hard to answer: If a creature like a comb jelly with a nervous system and muscles were the most ancient animal, what had happened that made the more amorphous-looking sponge give up all the biological bling? After the genetic toolboxes for jazzy cells and tissues were lost, did the cnidarians—the jellyfish, corals, sea anemones, and siphonophores that came along afterward—have to re-create them from scratch? If so, how did they do it?

Soon, however, other scientists carefully reanalyzed the DNA and returned the sponges to the base of the tree of life. Many scientists felt that a sense of order had returned to our notion of evolution. The problems, the new analyses explained, stemmed from comb jelly genes that were just so different from those of other animals that they tripped up the computer algorithms used to analyze the thousands and thousands of data points in the study. When the researchers looked at the data carefully, those errors were corrected.

However, the reanalysis sparked a fascinating debate. Some scientists pointed out that complexity isn't such a simple thing. Our idea of what's complex might be biased by our point of view, based on the intricate biological systems we know from our own bodies. They say that both sponges and comb jellies are complex, but in ways that aren't fully appreciated, in part because they have been so understudied.

Regardless of which is older, both the sponge and the comb jelly are certainly related to all other animals by very, very long branches of the tree of life, branches that get hazy when we try to see just where they attach to the trunk. What is clear is that the more we learn about these animals from our deep past, the more we will understand about the gnarled evolutionary tree that supports all the animals in the world, including ourselves.

6

Robojelly

As I followed jellyfish for months and then years, I found that the smallest hint from the world around me might bring on visions of medusae, moments that were admittedly apropos of nothing for the people in my life, especially my kids. Driving the carpool to school, I might see a pigeon alight on a telephone wire and notice the graceful flap of its wing tips. From my chauffeur position in the front I'd say something like, "Did you know jellyfish are the most efficient swimmers in the world because of the flap on the end of their bell?"

From the backseat, Ben would answer, "Mom, could you please change the radio station to 94.7?"

Fiddling with the buttons, I'd try again. "Don't you think that's amazing?"

In the rearview mirror, two sets of uninterested lavender-trimmed green eyes replied that they didn't think it was amazing at all.

After school, as Isy sipped on a smoothie, I'd say, "Jellyfish actually suck themselves through the water, like the way you're using that straw." Her response would be a drawn-out enumeration of the rules of a playground game that was like tag, but not really.

Though I had tried, my kids did not feel the pull of jellyfish with me. Isy would occasionally draw pictures of jellyfish on the bottom of the whiteboard where I'd started to keep a list of jellyfish names in foreign languages, but Ben made it clear that unless they were on YouTube, jellyfish were not for him. The world of jellyfish was one that I swam in alone. And though a piece of me wanted to share it with them, another part luxuriated in the fact that it was all mine.

My start-and-stop life juggling freelancing and family meant that the bulk of my day focused on jobs that paid decently but were often mind-numbing. You know all those A, B, and D choices that you didn't pick in the problems at the end of every chapter in your science textbook? Someone had to write those, and that someone might have been me. And while my kids challenged me in so many wonderful ways, my activities as a mom didn't have much intellectual depth either. I certainly found no vicarious thrill relearning letters and numbers, re-memorizing grammar, or restudying spelling after my workday ended and their homework began. As my kids progressed through elementary school, jellyfish had become my own alternative school, a place where my mind could play, and explore, and expand, the intellectual playground that I craved.

Woods Hole, Massachusetts, situated on the southern edge of Cape Cod, is one of the crown jewels of ocean research, home to two major research institutions, the Marine Biological Laboratory (MBL), affiliated with the University of Chicago, and the Woods Hole Oceanographic Institution, affiliated with MIT, fondly called WHOI (pronounced whoo-ee). Summer brings ocean scientists from around the world to this scientifically vibrant village, a place where the proximity of other researchers and the ocean itself creates a fertile brew for innovation. After I returned from my study-abroad program in Israel, smitten with the ocean, I landed a summer job as an undergraduate intern in a marine ecology course for graduate students at MBL. Another summer, in grad school, I worked at WHOI.

Twenty years later, friends who lived near Woods Hole invited our family for a visit. If the Texas summer heat does have one thing to recommend it, it's the fact that it makes it easy to decide to go somewhere else. Especially if that place is near a cool beach. As had now become my habit, I took a day from our family trip to explore the nearby jellyfish science while Keith and the kids went on an alternative adventure. This time their destination was the putt-putt place with tinted waterfalls and a trampoline park next door.

During the first summer I worked in Woods Hole, I heard a guest lecturer speak. I don't remember his name, but I do remember that his topic was the physical tricks that ocean creatures use to withstand their wave-swept world. He explained that the dome shape of snails called limpets balances lift and drag on their shells so the pounding of a wave can't dislodge them from the rocks they cling to. He talked about how seaweed has found a different solution for withstanding the force of waves. It has evolved the flexibility and strength to move with and not against the fluid flow. The lecturer mentioned a book called *Life in Moving Fluids* by a scientist named Steven Vogel. I read the book thoroughly, absorbing its ideas and equations like a sponge. I was captivated by the connection between the natural world and the physical one, how the twin pressures of physics and genetics mold every creature on this planet. Physical forces are always making rules, forcing conformity, setting the boundaries. Gene mutation is always seeking diversity, coming up with innovation, pushing boundaries. Between the two are the multitude of brilliant solutions to life that we find on this planet. In biomechanics, I thought I'd found life's calling. I hadn't, but I still have the dog-eared copy of Vogel's book on my shelf, a talisman from a career that never happened.

The scientists I'd arranged to meet in Woods Hole that day study the biomechanics of jellyfish, which might seem a poor proposition. Jellyfish are often called drifters, creatures swept along in the currents, wanderers that can't control their movement. But that's far from the truth. These scientists have learned that jellyfish are some of the best swimmers in the world.

Driving down the tree-lined road that leads to Woods Hole, I was grateful for the solitude, for being able to feel the flood of my old memories alone. As my rental car tipped forward over the hill on the edge of town, a tidal pond came into view, adorned for summer with sailboats swinging on their moorings. Its name, Eel Pond, I remembered right away. To the right, I saw the bar where my friends tricked the bouncer so I could sneak in when I was underage. Once, one of them grabbed the poor guy by the shirt and dragged him over to the drawbridge separating Eel Pond from the open sea of Vineyard Sound. She clamored, "A great white! I just saw a great white go into the pond!" While the bouncer investigated the alleged shark sighting, I slipped inside and hid among the scientists and graduate students, who filled and refilled my beer cup. Another night, we'd all sat around a campfire on the beach when a local police officer approached to tell us that fires were illegal. The international students in the class broke into their native tongues, a spontaneous Tower of Babel that sent the officer into retreat. We'd kept the fire going until the sun peeked over the horizon.

Looking for parking, I recognized the trail to the rocky beach where I'd gone snorkeling with a sandy-haired surfer, an ichthyologist who effortlessly recalled the scientific names of every fish we saw. His Latin words sounded like poetry to me. I'd fallen helplessly in love. I looked up and saw the gray-shingled lab building where very late at night we drank gin and tonics from Erlenmeyer flasks, pushing white pickled onions down their narrow necks. I remembered the acrid smell of the lab, the sound of bubbling seawater, the holes burned in the front of all my T-shirts from the hydrochloric acid used to wash the lab's glassware. Nearby, I saw the long jetty where that sandy-haired ichthyologist and I had the first of what were to be many horrible fights. Where he told me my love was one-sided and, to prove it, threw the keys to my dorm room deep into the ocean. The morning he left, I found a copy of *Life in Moving Fluids* propped against my dorm-room door. The inside cover was inscribed in neat handwriting: "I'll miss you tons." The emotionless words only shattered my heart again.

Maybe I should have expected the sentiments that overwhelmed me as I drove over that drawbridge into Woods Hole. A part of me had always thought that this would be my life, leaving my lab to meet my scientist husband for an afternoon beer before retrieving the kids from the local science summer camp. Was this whole jellyfish thing a way to retrace my past? To play at being a scientist again? Had I chased a dream back here?

Jack Costello, a professor at Providence College, and Sean Colin, from Roger Williams University, have had long careers studying the way animals move in the sea. Both are clean-cut, with hair close-cropped to the point of baldness and with healthy tans from the summer sun. Sean is a bit more reserved than Jack, but like the best of collaborators, they often complete or clarify each other's thoughts. It's a real joy to sit and talk to these scientists in their third-floor lab with the knockout view of Vineyard Sound. The conversation always starts off with jellyfish, but then it has a way of wiggling into places you never expect, as the best conversations do.

The project Jack and Sean were working on when I first met them was a grant from the U.S. Navy to build a robotic jellyfish. Sean said, "From the Navy's point of view, a robotic jellyfish could be a vehicle that can sit out in embayments for months. That's what jellies do. They can maintain themselves, often for months, in isolated areas. They have limited swimming abilities and tools, but the design is efficient, so they can be self-sustaining and could be used for naval surveillance."

I imagined secret code names for medusan spies: JellyOps, Agent Transparent, Swarm Team.

Sean and Jack weren't so interested in the spy side of the project, though. Sean said, "Our motivation to get involved was to, hopefully, learn from the process. And that's what happened. Within two years we had a functioning robot. And we immediately learned something very significant from it, from the very first time we worked with it."

The robotic jellyfish was fabricated by a collaborator named Alex Villanueva, an engineering student from Virginia Polytechnic Institute whose research expertise was vehicle design and propulsion. The droid, named Robojelly, was about the size and weight of a saucer. Alex used a 3-D printer to create a mold in the shape of a moon jelly's bell. He poured silicone into the mold, forming a flexible structure that moved like a jellyfish. Alex also developed a composite of spring steel and a special type of metal called a shape-memory alloy, which—like a memory-foam mattress—always returns to its original shape after being deformed. He formed the composite into eight arms splayed out from the center of the silicone bell like the spokes on an umbrella. When the electricity was turned on, the composite bent inward with a curvature similar to a real moon jelly. The shape memory alloy returned the jellyfish to its open position when the electricity turned off. The wires that juiced up Robojelly sprouted out of the top of the bell.

"It moved very similarly to a jellyfish, contracting and expanding just like a jellyfish," said Jack, explaining their first encounter with Robojelly. "So we put it in the water, turned it on, and it actuated." ("Actuated" is biomechanics talk for "contracting.") "And it was just doing this." Jack cupped his hand like a medusa. He bent his fingers inward and scooted his hand upward. But instead of continuing on the forward path, when he expanding his fingers, he lowered his hand to its original location. "And we were, you know, discouraged." Jack continued to mimic the up-and-down movement of the droid. Like a yo-yo. This robot jellyfish was going nowhere.

But then Alex said he had another trick up his sleeve. He had made silicone flaps that he could attach to the outside edge of the bell. He just hadn't had time to attach them. The researchers pulled Robojelly from the tank and made the changes.

They tested it again. "And immediately it was like . . ." Jack's hand imitated a jellyfish swimming off into the distance.

"That was a eureka moment," Sean said. Without the flexible flap, it turns out, a jellyfish can't swim. The pretty peplum that billows when a jellyfish pumps isn't there just to increase its elegance for our admiration. It's there because it drives the jellyfish forward. Alex later calculated that adding the flap increased Robojelly's thrust more than 1,000 percent.

Scientists don't call it the flexible flap, of course. They give it a more official-sounding name: passive margin. They suspected that jellyfish weren't the only creatures capitalizing on the benefits of these flaps. Jack and Sean, along with some colleagues, studied videos of fifty-nine different animals that swim or fly. No matter what the moving parts of the animals were made of—feathers, skin, or scales—or whether they were moving through water or air, all had evolved passive margins. Throughout the animal kingdom, from bugs to bats to birds to swimming snails and whales, the zoologists found passive margins. In whales it was the tip of the fluke; in birds, the ends of their wings; in fish, the edges of their fins. Furthermore, the passive margin is even built the same way in wildly different animals. Measuring from the base of the fluke or wing or fin to its tip, the bend begins just about two thirds of the way to the end, and the angle of bend is always between 15 and 40 degrees. At the risk of sounding risqué, neither a floppier flap nor a stiffer strut worked better. What makes the passive margin so fabulous is that it's passive; it just flaps as a consequence of the movement of whatever part it's attached to, be it wing or flipper or jellyfish bell. It doesn't require any muscle power; it doesn't require any energy input. (In movement on land, friction with the solid ground dominates the forward push, so we don't have flaps.)

Remember the mesoglea, the watery center of the jellyfish, which Monty Graham pointed out in the Discovery Channel documentary? Like a water balloon regaining its shape after it's been squeezed, the watery center of a jellyfish automatically springs back to its relaxed position without requiring any energy input. Couple the mesoglea's natural springiness with the passive margin's thrust, and you've got the most efficient form of propulsion we know. Let me emphasize that: Jellyfish are the most efficient swimmers of any animal anyone has ever studied. One of Jack and Sean's collaborators, Brad Gemmell, calculated that a moon jellyfish uses muscle power for only 20 percent of its motion. That's as if you went running but had to put out energy only every fifth stride. Salmon, among the most efficient swimmers of all fish, require three and half times more energy per pound to move the same distance. Animals that fly or run require a hundredfold more energy. "They aren't fast, but they move big distances with very little cost," said Jack. And to think, jellyfish have a reputation as drifters.

In a press release on the findings, Jack pointed out that engineers should take a tip from biology: "Flying and swimming animals have a much lower cost of transport than present manmade designs. That is part of our motivation for understanding biological design: Animals do it better." If you think about it, they should. They have had the twin pressures of genetics and physics working on them for millennia. That's a lot of time for mistakes to be corrected.

About a year after I spoke to Sean and Jack about Robojelly, Alex Villanueva, the machine's fabricator, unveiled Robojelly 2.0, a life-size droid called Cyro. It's modeled after the lion's mane, one of the biggest jellyfish, which can grow larger than my husband, who is around five feet, seven inches and 140 pounds. This swimming robot could carry a payload of sensors to collect, store, analyze, and communicate information about its surroundings. It could also carry its own rechargeable battery pack. Researchers at the University of Texas at Dallas were working on switching out that battery pack for a hydrogen power source that could derive energy from the seawater. If the work is successful, jellyfish robots really could monitor the seas for months on end. But despite the progress, jellyfish robots still have a significant design flaw when it comes to mimicking the real thing. Like the original Robojelly, Cyro has actuators that radiate out from the center of its body, while a real jellyfish has circular muscles that surround the edge of its bell like a belt. Alex told me that creating a structure that bends can be done, but creating one that contracts enough to squeeze the water out from the underside of the bell isn't possible right now. Trying to mimic the muscles of a creature from the primordial world, we creatures of the twenty-first century have bumped against the edge of our technological abilities.

At the other end of the size spectrum from Cyro, researchers at the California Institute of Technology and Harvard University have engineered a jellyfish just one millimeter in size. Instead of metal actuators, the movement of this artificial jellyfish is powered by biological muscles. The creation, called a medusoid, was developed by casting silicon in the shape of a baby jellyfish ephyra. The scientists etched a protein called fibronectin, which encourages cell growth, on the silicone in a pattern like that of the muscle fibers in an ephyra. They seeded the prepared silicone with immature rat muscle cells, which grew into the pattern sketched out by the fibronectin. When the researchers put the medusoid into a Petri dish and pulsed an electric current through the water, the medusoid fluttered with a motion that looked strikingly like a real baby jellyfish, and hauntingly like a heartbeat. That isn't a coincidence. Besides helping scientists understand the physics of propulsion, the medusoid was created as a new way to conceive of regenerative medicine. Growing organs to replace damaged ones has long been an idea with great promise, but hearts created from living tissues so far have been too weak to pump blood throughout our bodies. The powerful twitch of the half-synthetic, half-natural medusoid is a new approach to the problem of growing hearts.

For all its ubiquity throughout the animal tree of life, the flexible flap isn't even the most interesting story jellyfish have taught us about life in motion. When we think about moving, we think about pushing back against something. When we walk, we push against the ground to move forward. When we swim, we push the water backward with our arms to move ourselves forward. That push creates a zone of high pressure behind you, which propels you forward. Jellyfish do that when they squeeze their bell to push out water, but that's not all they do or even the most important thing they do.

One thing you'll notice looking at animals swimming is that they wiggle. Fish and sea snakes wiggle side to side. Whales and dolphin wiggle up and down. The theory has been that the wiggle creates little pockets of high pressure behind the animal, and those little pockets push the animal forward. But the jellyfish taught Sean and Jack that that theory was completely backward.

Working with Brad Gemmell and John Dabiri at Caltech, Jack and Sean developed a high-tech laser system coupled with super-high-speed video that allows them to see exactly how the water moves around animals. The animal is illuminated with a thin flat sheet of laser light. The water is seeded with very small beads that reflect the light. Each bead is neutrally buoyant and so moves like a tiny parcel of water. The high-speed video records the movement of every bead, which is fed into a computer and converted to hundreds of thousands of measurements of velocity. A computer algorithm converts velocity to pressure. The result is like a weather map, but for pressure on a scale of centimeters.

When the scientists put a moon jelly in the laser video system, what they saw upended everything we used to think we knew about swimming. Yes, the jellyfish created a region of high pressure behind it, but that was only part of the story. More noticeable, and never before seen, was a region of low pressure near the topside of the bell. The low-pressure region was significantly bigger and more important than the high-pressure one. The jellyfish was not only creating forces to push it forward from underneath as it bent its bell; it was also generating suction to pull it forward from the top, much as you use low pressure in your chest cavity to suck in air. Did you just suck in air? I hope so, because this discovery is astonishing.

Sean and Jack and their colleagues discovered that animals bend not so much to *push* themselves through the water but more to *pull* themselves through the water. Bending creates rotation—eddies, little whirlpools, or vortices—in the fluid. At each vortex's core is a cylindrical region of low pressure. When these vortices merge, they create a whole region of low pressure that can suck an animal through the water.

Jack drove the point home: "The reason an animal bends is not for any other reason, really, than to create those low-pressure regions around its body, so that its body is pulled, or sucked, into that space. Otherwise, an animal might as well be stiff."

Noting that for all of human history, the vehicles we've built for going through water, things like boats and submarines, are indeed rigid, Sean said, "If you go back and look at fluid dynamics, this assumption of pushing off the water has underlain so much of the work for such a long time. It's affected the way we've designed underwater vehicles. Yet it's probably much more efficient to pull yourself through the water than push." From our myopic vantage point on land, where we push against the ground to move, we never guessed how important it could be to create a space to pull us through the water.

"Do you think it's odd that jellyfish, what might be the oldest animal to swim, found this optimal bending solution? This way of slipping through the water?" I asked.

"I think you should turn the question around," Jack answered. "There were probably a lot of primitive animals that went extinct because they didn't find this solution. The only ones we get to see are the ones that did it best."

Leaving the biomechanics lab, I walked into the foggy evening air. For old times' sake, I strolled around the campus, past the trail to the beach where we used to go night snorkeling and dance around campfires, past the lecture hall where I often made pot after pot of coffee to keep grad students awake, and past the jetty where I'd cried heartbroken tears. I spotted a group of students gathering outside the dining hall. They were joking and jostling. Some were carrying towels and beach gear; others were toting coolers. I knew they were planning to find a remote spot to hang out, drink, and talk science. They are part of a tradition of summer students who claim this piece of Cape Cod as their training ground and who, while they live here, believe they own it.

I turned away from them and walked around the rear of one of the lab buildings. I noticed a piece of art that had been installed since I was here last; I walked toward it. On a set of three stone stairs, the words SEE, YOUR, SELF. I climbed the steps. On the top step, I was face-to-face with a human-size panel, a photograph of a person wearing a Marine Biological Laboratory T-shirt and a baseball cap. But in place of the face was a mirror. Down in the grass at the bottom of the stairs, I saw the granite placard: *Relaxing Nobel Prize Winner* by David Bakalar. I considered the reflection of myself, posing as a scientist. The same curious caramel eyes specked with dark brown that had looked back at me my whole life; the same deep black pupils stared back. But now, hints of gray at the part in my hair, thick indentations around my mouth, cheeks starting to sag.

Nope, I thought. I'm not a scientist who goes blue-water diving for effervescent jellyfish or assembles lasers and high-speed video cameras. I'm never going to be the scientist who studies the biomechanics of jellyfish or any marine organism. I won't discover any of the groundbreaking answers about the nature of swimming. I've left those dreams to others.

I turned and walked down the stairs, slipping into the space I'd begun to create for myself.

Seeing What's Not There

The FedEx tracking e-mail announced their arrival at 11:23 a.m. Sure enough, when I opened my front door, I found a small cardboard box holding a container with nine pounds of water and a few grams of gelatinous pets. After my trip to San Francisco, when I met Alex Andon of Jellyfish Art, I had ordered my own jellyfish tank from the company's hugely successful Kickstarter campaign. I was one of the supporters who helped raise almost $600,000 from preorders. It had taken more than a year, but I'd finally received my elegant Danish-style acrylic tank. I had added seawater and gravel, as instructed, and turned on the circulating motor. The white sound of water bubbling harmonized with the hums of the refrigerator in the kitchen and the whoosh of the air-conditioning in the dining room. The only thing missing had been jellyfish. But now they'd arrived.

I opened the FedEx box, cleared away the packing popcorn, and tore open the bubble-wrap bag. The top of a tightly twisted plastic bag secured with a rubber band poked out. I snipped away at the band and unraveled the coil. At first I saw nothing. Then slowly, pulsing soundlessly, like transparent fairies, three small moon jellies appeared—pure life among so much synthetic material.

I began transferring water from the bag to my aquarium and from my aquarium to the bag, so the jellies could adjust to the new chemistry that would surround them. The process, which seemed simple enough, was made nerve-racking by the fact that I could barely see the jellies. On the eighth transfer, I caught the lip of one jellyfish in my cup. It slid backward like a blob of mucus. Yikes? After the twelfth transfer, I placed the plastic bag inside the tank, turned it sideways, and slipped the jellyfish out. There was a tense moment when I worried that one had been caught up in the plastic. It was so hard to tell transparent from transparent.

While I was in Woods Hole, I had met with a jellyfish scientist named Richard Harbison, who had helped pioneer blue-water diving and research on jellies from underwater vehicles. In the course of the interview, I had asked him why so many jellyfish were transparent. He looked at me as if I'd asked him why grasshoppers were green. "The question is, why isn't *everything* transparent?"

I'd never thought about it before, but once I did, it made as much sense as a grasshopper in green grass. In the ocean, with nothing to hide behind and nothing to blend into, being opaque puts you at risk of being seen. Being seen means being eaten. Invisibility is also an advantage if you are a predator. Prey are more likely to steer clear if they can see you. Prey might just stumble into your grasp if they can't. Transparency is an underwater invisibility cloak; the ultimate disguise.

As I walked through my house in the days after the tank was first set up, I'd find myself stopping in front of it, meditating on my jellies, on their insane transparency. It's tempting to think that when we look at the world, we see what's in front of us, but in reality we can see only what our eyes allow us to see.

I had fallen head over heels in love with marine biology at Woods Hole (and head over heels in love with the ichthyologist surfer, too), but I followed many of my friends to Boston after graduation. Although seeing the world through equations didn't come naturally to me, I used my skills to my advantage. I parlayed my degree in math into a decent-paying job as an accountant and found a cute second-floor walk-up on Bunker Hill to share with a friend. About eight months later, my roommate came home from work to find me curled miserably on the floor. Accounting wasn't in me.

"Go to grad school," she said. And I knew she was right.

Google didn't exist yet, so I went to the Boston Public Library's reference room and found a volume of all the country's universities and faculty. I made a list of fifty marine biology professors and wrote each of them a letter on cream-colored stationery, describing my passion for marine science and my brief experience at marine labs, and I didn't fail to flaunt my math degree. I was trying to appeal to biologists who either loved math or hated it. A lot of biologists, I'd already discovered, have a fear of math, so I was hoping one might want *me* to do it instead.

If Sönke Johnsen had been teaching when I sat down at the Boston Public Library, I would definitely have sent a cream-colored letter to him. A scientist at Duke University, he works on the way animals in the sea interact with light. Sönke's work straddles the messy complexities of the organic world and the predictable organization of the physical world. Like me, Sönke found his way to this scientific outpost by way of an undergraduate degree in mathematics, and his work is peppered with the scrolls and swirls of equations.

When I posed it to him, Sönke told me that the best answer to Richard Harbison's question of why everything isn't transparent is: because it's hard.

In order to be transparent, every part of you has to be clear, from the outside right on through. Most animals that live near any surface are not transparent. The surface they live on provides a backdrop to blend into, which means they only have to decorate their surfaces to disappear. A zebra's stripes are only in its hide, and a flounder is mottled only as deep as its scales. Transparency is much more than a surface paint job.

One way animals can maximize transparency is by minimizing the number of cells they have. Some animals do this by being extraordinarily thin. Light passes through such animals so easily that it encounters almost nothing to scatter it. One supermodel of transparency is a comb jelly called the Venus girdle, which is less than an inch thick, but can stretch a yard and a half in length. The animal's svelteness decreases scattering so much that in photographs it looks like just a few white lines dashed off by a sketch artist. A lot of baby animals are sent out into the ocean swathed in the transparency of flatness too. Young eels are so skinny and clear that if you laid them over this page, you could read the words right through their bodies.

But many jellyfish aren't flat; they are bulbous round things. Or are they? Sönke said that in a lot of ways, jellyfish are two-dimensional animals made three-dimensional by the colorless jelly they are named for. Remember the mesoglea, the watery jelly inside the jellyfish? The baritone narrator of *Rise of the Jellyfish* could have added one more mesoglea advantage to his list. It not only gives the jellyfish its swimming recovery stroke for free, not only offers buoyancy, and not only works like a scuba tank for survival in polluted water, but also provides an unbeatable disguise.

Because it's so watery, light passes through the mesoglea nearly the same way it passes through the ocean. Watery mesoglea is a brilliant solution to being transparent: Jellyfish are physically fat and optically skinny. If I were to shine a flashlight through my pet jellyfish, half or more of the light I would see would pass right through, depending on the color. And being physically rotund matters a lot in a place like the open ocean, where there's nowhere to hide, because even the best invisibility cloak isn't perfect.

Those thin outer and inner cell layers, skinny as they are, do reflect some light, especially in deeper, darkened seas. If an animal has the bad luck of being spotted by a predator, it's always better to have heft on its side. No one wants to tangle with a big dude. The mesoglea allows a thin jellyfish to masquerade as a jumbo version of itself, and it does so on the cheap. Because mesoglea is made mostly of water, it is a form of life that is not cellular. Like our hair or our fingernails, mesoglea is not nearly as metabolically costly to maintain as muscle or nerve cells. So jellyfish get a lot of heft, not for free but at a significant metabolic discount.

The movie *Finding Nemo* did a good job of representing the hallucinatory world of the deep sea. Part of the reason it succeeded was that its creators asked people who'd been there what it looks like. Sönke was one of those people. He has been fortunate to make over a hundred dives in submersibles to depths where the light runs out. He has written about descending into the ocean to a depth of three hundred feet, the length of a football field. As you go down, the light around you gets less and less vibrant. At the fifteen-yard line, the water molecules have already sucked out all the red light. If you'd painted your fingernails red and happened to look down at your hands, your nails would look jet black. At the fifty-yard line, the yellow light becomes extinguished, turning blondes into brunettes without using dye. Ten yards later, much of the green disappears. At a depth of eighty yards, the world seems lit by a single blue spotlight.

When we flood deepwater animals with light from a camera flash, with more colors than they would ever experience in their lives, many are red as Santa Claus's suit. The blood belly comb jelly, which lives a half mile deep, is a deep currant color, especially its stomach. The jellyfish named big red (*Tiburonia granrojo*) is a stocky creature found in the Pacific at depths of a mile or more. Just discovered in 2003, it has a warty body that may reach a yard across, and four to seven thick arms that look like slugs emerging from under its bulbous bell. The whole creature is the color of a rich cabernet. A jellyfish called *Atolla wyvillei* trails about two dozen tentacles that look like they were soaked in merlot.

"But why wouldn't they just be transparent, like moon jellyfish?" I asked Sönke. Clear certainly worked well for my pets. I could hardly see them in the tank.

Red is even better than clear at depth, Sönke explained. Clear can reflect stray light, especially around the edges. Burglars wear black because it absorbs every color of light. Nothing reflects off black, making it hard to spot. In the deep ocean, where there is no red light, the color red is as effective a camouflage as black. While vertebrates can and do manufacture black, it's thought to be easier for invertebrates to make red pigments. For all these deep-sea animals, "red is the new black," Sönke told me.

That first year after college, I wrote dozens of letters to marine-science professors across the country. Finally one wrote back. He had half of a research assistantship to offer. Half. I jumped at the chance to live in one of the most expensive cities in the United States on a salary just a sliver above the poverty line. The job was in Los Angeles, and lo and behold, so was the ichthyologist surfer.

The apartment I could sort of afford was just north of downtown Watts, where Rodney King would soon be cruelly beaten by the police, and was within walking distance from the University of Southern California, where I would be working. The place came furnished with a roommate who had a scruffy little dog that peed on the floor every time I walked in the front door and a kitchen with more cockroaches than edible items. I chose to see none of that. I saw, instead, the elegant sweep of the avocado tree outside my bedroom window, already laden with green fruit. I walked to the lab (a lab!) focused on the warm sun and the pale-blue cloud-free skies. My view was graffiti-covered concrete rather than blue, sparkling sea, but I knew I could get to the ocean just a few miles to the west.

When I wasn't accepted to any graduate school program to study invertebrate biomechanics, I had to reckon with the twin pressures that mold so many of our lives: passion and economics. Passion pushes our curiosity, drives innovation, and breaks through boundaries. Economics makes rules, sets boundaries, and forces compromise. During my half internship, I met a professor of marine biology who studied photosynthesis and the ecology of plankton, using mathematics to simplify the complexities of both. He was brilliant and kind, and our personalities clicked. He offered me a spot as a graduate student in his lab. I convinced myself that single-cell plankton were almost invertebrates and that photosynthesis was almost biomechanics. Plus, I could stay in L.A., near the ichthyologist surfer. He still didn't love me, but he did tolerate my attempts to be his girlfriend.

We've all been in a place where our emotions take over our view, coloring our visual field so that we see what we want to see, and that was certainly the case for me in L.A. Soon enough, I'd discover that the clear skies were a symptom of a grueling drought, that the avocados that grew on the tree outside my apartment were tough and inedible, and that the traffic west to the beach was bumper-to-bumper, always.

During one of my final years of graduate school, a red tide—a bloom of a kind of phytoplankton that absorbs green light, tinting the ocean crimson—spread along the shores of the Pacific Ocean more than five hundred miles, from Monterey, California, down to Baja California and forty miles offshore to the Channel Islands. It was the biggest red tide California had seen since 1902. I'd moved to a slightly run-down beach cottage. It was a sketchy proposition to use the kitchen sink disposal unit, but in an improvement over my earlier residences in L.A., it was mostly cockroach-free. The little house's most redeeming features were a fragrant vine of jasmine running up the side of the front door and, if you stood on tiptoe, a view of Hermosa Beach, which was smack-dab in the center of the red tide. One night, at its height, I bundled up against the winter desert air and the ocean's biting wind and walked down to see the spectacle.

Offshore, a wave rose up like a three-foot-high black wall. My eye was drawn to a small section of the top of the wall as it tipped toward the land. Just there, a pale electric-blue light gurgled, seemed to be held in place for a moment, and then like a neon-blue zipper streaked along the breaking edge. Barreling toward shore, the blue infiltrated the entire wave front with a wild cloud of incandescence. As the wave exhausted itself, the bioluminescence swirled and curdled in the diminishing space between the wave and the slosh, like blue glowing embers burning out in a watery hearth. It was breathtaking. The creatures responsible for this light show were single-cell flagellated phytoplankton called *Lingulodinium polyedrum*, and as I was now studying phytoplankton and light in grad school, I had brought an empty jam jar to fill with the glittering seawater. Days later, in my dusty dark lab far from the sea, I gave the jar a swirl. A glowing blue whirlpool reignited the water. In the ocean, especially below the blue laser depths where sunlight is extinguished, it's thought that 90 percent of the creatures glow. Creating light is one of the commonest ways in which marine organisms communicate with one another. When I swirled those phytoplankton in the jam jar, I was seeing their screams.

Sönke Johnsen received his PhD working in the lab of Edie Widder. While Sönke talked to me about what happens to light that already exists in the sea, Edie's work focuses on the creation of light by living organisms: bioluminescence. Edie began studying how phytoplankton like those in my jam jar create light, but shifted to deep-sea animals when she got the chance to dive in a single-person submersible. Like Sönke, she descended to the depth where sunlight runs out, then turned off the operating lights on her submersible. Suddenly the ocean exploded in so much light she could see the numbers on the instrument panel dials. She thought an electrical short had caused a burst of electricity. But it wasn't the instrumentation. It was a chain of jellyfish, a thirty-foot-long siphonophore that had bumped into the submersible and shouted its surprise with light.

When Edie drives a submersible through the deep, the animals hit her windshield and erupt in squiggles and blasts of light that look like what the Star Wars *Millennium Falcon* experiences when accelerating into deep space. In the Gulf of Maine, where she has worked often, Edie can name each animal by the light it produces: the hops of copepods, the bursts of comb jellies, the sparks of shrimp. Just as a visitor to a foreign country gleans meaning from a language she doesn't fully understand, Edie has become an interpreter of light signals in the deep sea. Like so many important conversations on land, she told me during a phone call, nearly all of it has to do with sex or food.

There's a sea cucumber that says, "You won't get away with this," when it releases a sticky bioluminescent goo. Like the dye pack enclosed in a bag of stolen money, the goo marks the attacker, making its trespass known and signaling its position to any lurking predator. In the dark sea, a creature marked with light is soon someone else's dinner. There's a copepod that says, "Fooled you!" when it releases bioluminescent globs of light from the glands on the back of its legs. In the same way a jet fighter releases decoy flares to lead a heat-seeking missile astray, the copepod's light globs are meant to put a predator off its scent. Several species of squid and fish say silently, "Nothing to see here," when they illuminate their undersides with the same intensity as the dim light coming from above. The counterillumination obliterates the animal's shadow, making it invisible to both predators and prey. A jellyfish that lives deep in Monterey Canyon says, "Go fetch!" It has thirty-two bioluminescent tentacles that it can shed at will, tossing a glowing lure into the distance. The attacker ends up with a single mouthful rather than an entire meal.

The sexiest bioluminescence I know is made by an animal called an ostracod, a sesame seed–size crustacean that lives inside a hinged shell. The male ostracod asks, *"Voulez-vous coucher avec moi?"* but in a language that only female ostracods of the same species can understand. About fifty species of ostracod are bioluminescent, and each one has its own pattern of Morse code–like dabs of light that it paints in the water. Some release a vertical line of glowing spots, spaced closer and closer together as the animal climbs in the water. Some release vertical dabs in the opposite direction, moving downward. Some release dabs on the diagonal; others, the horizontal. Some males form teams, releasing diamond-shaped dab clusters. Some do a bit of a box step, a jog to the right and then a dash upward. Each is a glowing love letter recognizable only to that special female of the same species.

Anglerfish have long rods extending from the top of their head. The end of the rod is lit with a glowing bulb and dangles in front of their sharp teeth. To shrimp or small fish looking for a snack, the light calls out, "Come and get it!" like a camp chef clanging a soup pot. When animals come looking for a plankton meal, they instead become a meal for the anglerfish. A literally brilliant double entendre, the lure also says, "Come and get it . . ." to potential mates. In anglerfish, the males are just a fraction of the size of the lure-bearing females. For them, dating is deadly serious. Choose the wrong lady, and the tiny male is dinner. So a female anglerfish's lures speak directly to a mate of her own species. Only he will recognize its shape or the frills she decorates it with. After he's spotted the right mate, a male anglerfish sneaks around her belly and with one fateful kiss embeds himself there, forever fusing his tissues with hers. Once they're conjoined, the female feeds the male by joining her bloodstream with his. She builds passageways into his body so that she can harvest his sperm when she's ready. And her love light isn't extinguished just because she's mated with one male; she can have several suitors attached at the hip.

One of the most common shouts of light from the dark sea is "Help!" The call is a burglar alarm, a wild and obtrusive visual siren. The flare requires a lot of energy, so an animal will use it only when its life is in danger. Its goal is to get other animals to notice an attack under way, in hopes that one of them will come to the rescue—not to save the animal giving off the light, but to make a meal of the attacker. The bioluminescent version of "Help!" is what the phytoplankton in my jar were saying with their bright blue sparks, and the cry comes in many forms from different species. A blast of light is common. A number of deep-sea shrimp emit a cloud of blue light to cry "Help!" and perhaps "Boo!" too. On a submersible dive off of Africa, Edie recovered a jellyfish that, like the car alarm that won't shut off in the mall parking lot, has a call for help you can't ignore. Like so many deep-sea animals, deep red *Atolla wyvillei* has several dozen tentacles that look like strands of crimson yarn. On the ship, Edie let it acclimate to an aquarium in a darkened room. When she touched it just on the edge of its bell, the animal lit up like the Wheel of Fortune. Bulbs of light erupted at the point of contact and then cascaded clockwise around the entire bell, sending short sunbursts away from the center. The circling and sunbursting continued around and around the bell for nearly a minute. It was a call for help like none other. And it gave Edie an idea.

During her years in the dark water watching animals, Edie had started to get the feeling that we'd been going at undersea research all wrong. Nets shred delicate gelatinous animals, giving us a biased view of the sea. "I defy you to find any other branch of science that still depends on hundreds-of-years-old technology," she said. While submersibles are an improvement, she'd piloted enough different types of underwater vehicles to notice that she'd seen more animals in models with quieter electronic thrusters than in those with noisier hydraulic controls. Loud, bright vehicles seem to scare away more skittish animals. Scientists had tried leaving dead bait on the seafloor and videotaping what came to eat it, but those lures brought in scavengers rather than animals that hunt live prey. Edie wanted to say, "Come and get it!" to the animals of the deep. The bioluminescent signal of *Atolla* would be that message.

No granting agency would give Edie money for a project whose outcome wasn't certain. So she cobbled together bits and pieces of equipment to build an electronic bioluminescent jellyfish, which she called e-jelly. You can see the word "Ziploc" on its topside, because the housing was made from the same kind of round plastic container you might use to store leftovers in your refrigerator. When the e-jelly was turned on, blue lights circled around and around the edge of the container. Edie connected the bioluminescent lure to a red spotlight and a video camera that detected red light. Because most undersea animals can't see red, they would be oblivious to the fact they were being filmed.

Edie first deployed the e-jelly in the Gulf of Mexico in 2004. She dropped the system on the edge of an underwater lake of super-salty water where methane naturally bubbles up to the surface and feeds a deepwater ecosystem of bacteria, mussels, and clams. She turned on the video but left the e-jelly dark. The footage showed fish swimming by lazily. Edie was thrilled that they didn't appear frightened. She said, "I had my window into the deep sea. I could, for the first time, see what animals were doing down there when we weren't down there disturbing them."

After four hours of deployment, the e-jelly turned on its light show. And just eighty-six seconds after it began to shout "Help!" the camera's viewfinder filled with something no one had ever seen before. The sleek, smooth tentacles of a six-foot-long squid filled the screen. Its oval head and suckered arms extended over the e-jelly, as if to grab something nearby. And then, catching only water, the animal retreated out of the frame. The e-jelly had done a spectacular job. When Edie tried to identify the squid, she discovered it couldn't even be placed in an existing scientific family.

Edie's e-jelly caught the attention of several scientists who were on a squid hunt of their own. They wanted to find the largest squid in the world, *Architeuthis*, the legendary sea monsters that attacked the good ship *Nautilus* in Jules Verne's 1870 novel *Twenty Thousand Leagues Under the Sea*. We knew they existed. Fishermen occasionally encountered dead giant squid the length of a two-story building. Giant sucker marks are regularly seen on the skin of whales. But a live kraken had evaded detection for all of human history. After hearing Edie's story, the squid hunters thought that perhaps they'd been going about it all wrong. Perhaps Jules Verne's beasts were instead timid monsters. Perhaps they needed a lure that softly spoke squid.

Edie and her e-jelly were invited on a Discovery Channel expedition to Japan, where they landed the glowing lure on the bottom of the sea. The e-jelly made its "Come and get it!" call and, in an episode that is now truly legendary, a live giant squid replied. Six times the massive creature danced through the camera's view, as if flirting. The first four times a single sinewy arm or two reached into the frame, winding in and around the e-jelly housing. On the fifth approach, a single pointed tentacle crept into view from the left and then in a heroic burst, the entire animal launched itself over the e-jelly toward the camera, arms extended, its mouth in the center of an exuberant fan of tentacles and suckers. It's an incredible sight.

Every time scientists plunge into the depths, they glean a little more of the language and culture of that foreign place so close to us. Unfortunately, the submersible in which Sönke and Edie took many deep-sea voyages was sold to an oil exploration company in Brazil in 2011, ending nearly four decades of outstanding scientific research into the most unknown part of our planet. Funding to permanently deploy a long-term e-jelly mooring hasn't yet materialized. Edie, who was awarded a MacArthur Genius Grant in 2006 for her cutting-edge work in the deep sea, established the Ocean Research & Conservation Association, or ORCA, the first nonprofit foundation focused on using technology for conservation. She poured all of her award money into the foundation, but it's not enough. In 2013, funding for space exploration outpaced funding for the ocean exploration 150 to 1. Only three people have descended to the Marianas Trench, but more than five hundred have gone to space, and twelve have stepped on the moon. We need a NASA for the sea, Edie concluded at the end of a TED talk describing her search for the giant squid. Further, she has said, "more than 90 percent, 99 percent, of the living space on our planet is ocean. It's a magical place filled with breathtaking light shows, bizarre and wondrous creatures, alien life-forms that you don't have to travel to another planet to see."

She's so right. The nearest planet that may hold life is Mars, which is over 33 million miles away. A six-foot-long squid that we've only glimpsed swims less than a thousand miles away from me as I sit here in Texas right now. We know of about a quarter million species in the ocean, but a 2015 survey of the sea's creatures revealed that as many as 2 million more await discovery. What other great surprises are waiting for us in our own deep sea, if only we would look?

8

Day-glo Jellies

Just before Vesuvius shot glowing embers into the dark sky—their light a signal of impending doom—the Roman naturalist Pliny the Elder wrote, *"pulmone marino si confricetur lignum, ardere videtur adeo ut baculum ita præluceat,"* extolling the ability of a jellyfish to light the way like a torch when rubbed on a walking stick. I asked a jellyfish scientist if she believed that trick would work, and she was doubtful. But without electricity, the nights were darker for Pliny two thousand years ago.

A thousand and a half years later, when the world's greatest organizer, Carl Linnaeus, published the first scientific classification of animals, plants, and minerals, called *Systema Naturæ*, he concluded his description of the genus Medusa with these poetic words: "and most of them shine with great splendor in the water."

In 1870, Jules Verne wrote of an episode aboard the famed *Nautilus* in which the crew slipped into a swarm of bioluminescence:

> No; this was not the calm irradiation of our ordinary lightning. There was unusual life and vigour; this was truly living light! In reality, it was an infinite agglomeration of coloured infusoria, of veritable globules of diaphanous jelly, provided with a thread-like tentacle, and of which as many as twenty-five thousand have been

> counted in less than two cubic half-inches of water; and their light was increased by the glimmering peculiar to the medusæ, starfish, aurelia, and other phosphorescent zoophytes, impregnated by the grease of organic matter decomposed by the sea, and perhaps, by the mucus secreted by fish.

Jellyfish have long been known for their ability to glow, but the inner workings of the shine of one particular jellyfish—the mechanism behind the mysterious "grease of organic matter" to which Jules Verne refers—has had a truly luminous impact on all of our lives.

In 1945, sixteen-year-old Osamu Shimomura heard the familiar sound of air-raid sirens. Looking up, he saw a U.S. warplane heading south toward nearby Nagasaki. He watched it drop three parachutes from its belly. Relieved that the attack appeared limited, Shimomura returned to his work as an airplane mechanic. Suddenly, a bright flash filled the building where he worked, temporarily blinding him. Less than a minute later, a loud explosion and a strong pressure wave rolled across his eardrums. A dark cloud filled the sky. An atomic bomb had just released its fury only twelve miles away. The bomb exploded with 25 percent more force than the atomic bomb that had been dropped three days earlier on Hiroshima. In a flash, it killed forty thousand people and injured another forty thousand. It also brought about Japan's surrender six days later—the end of World War II.

Amid the destruction of both human life and infrastructure, it took three years for Shimomura to piece his life back together. He managed to land a position as a teaching assistant and then a visiting researcher at Nagoya University, where one day the head of his lab showed him a jar of dried ostracods, those crustaceans that write bioluminescent love letters in Morse code globs of light. The advisor removed a few and crushed them in his hand. Then he drizzled a bit of water on the powder. A blue glow shined from his palm. During the war, the Japanese army had used the glowing crustaceans as a battery-free light source—just add water. Shimomura's boss asked him to study the chemistry of the glow.

On the other side of the Pacific Ocean, scientists at Princeton had been trying to unravel the secret of bioluminescence for nearly forty years, but they hadn't progressed very far. They found that the glow required a chemical that they called luciferin and an enzyme to activate the chemical that they called luciferase. But the nature of luciferin was a mystery. Was it a protein? A protein connected to a sugar? A lipid? Something else? The first step to understanding the chemistry was to purify the luciferin. Shimomura knew that only a pure sample would form a crystal. Over ten months, he greatly increased the purity of the sample, but coming up with a pure crystal remained elusive. After months of unsuccessful experiments, Shimomura went home from the lab one night, leaving his sample in a solution of hydrochloric acid. When he returned, small red crystals had formed in the dish. Inadvertently, he'd done it.

After Shimomura's success, the head of the Princeton bioluminescence lab, Frank Johnson, invited the young chemist to the United States to continue his work on bioluminescence. They would need more material than could be harvested from tiny ostracods, so Johnson's plan was to study the glow of a jellyfish called *Aequorea victoria*. Three or four inches in diameter, the animal is crystal clear. The lower half of its bell is decorated with close-set white ribs that end in a fringe of thin tentacles. If you touch *Aequorea* in the dark, just where the ribs meet the tentacles, the bell shines with a bright green glow. (It will be important later to remember that the jellyfish's glow is green.)

Aequorea jellyfish were abundant at Friday Harbor Labs, which is roughly the equivalent to the Marine Biological Laboratory in Woods Hole, but on the opposite side of the continent, north of Puget Sound. Shimomura and Johnson arrived early in the summer and set to work collecting the animals off the marine institute's dock with nets that looked like pool skimmers. Using scissors, they dissected off the edge of the bell where the glow was concentrated, and squeezed the jelly through a cotton cloth. The liquid, which they called "squeezate," gave off a green glow for several hours. To purify the luciferin from the squeezate, Shimomura needed to inactivate the reaction. He tried salts, metals, enzymes—anything he could think of. But nothing worked. The squeezate kept glowing. Shimomura took long rows in a rowboat to meditate on the problem, often falling asleep under the warm summer sun.

I knew the dock that Shimomura stood on to scoop *Aequorea* jellyfish from the sea. I spent half a summer in the same spot where Shimomura struggled to discover the secret of the jellyfish's glow. After I was accepted to grad school, the department looked through my transcript and noticed that my biology background was uncomfortably thin. They suggested that I take a summer course in the basics, and I enrolled in a class on marine invertebrates at Friday Harbor Labs. It probably goes without saying that the ichthyologist surfer was taking a summer class there too.

The course I took was a survey of the bulk of the animals on Earth, twelve of the thirteen animal phyla, taught at breakneck speed twelve hours a day for six weeks. I still have the textbook, written by the course's teacher, Eugene Kozloff, or Koz, as all the students called him. It's warped with seawater spills and age, but I still refer to it when I can't remember something about some boneless creature I studied that summer. Koz taught by physical example. Every time he introduced a new group of animals, he'd walk us out to some sandbar or mudbank and instruct us to dig or scoop or flip something over. Then we'd gather around while he, gray-haired in chest-high green waders and wire-rim glasses, cradled in his palm some wriggly creature I'd never heard of before. He'd go on to explain precisely how this animal was an evolutionary success story, how it came to be, and why it was so special.

One day someone, maybe it was Koz, walked us down to the dock and pointed out a swarm of bulbous creatures in the water. It was a bloom of comb jellies, like the *Mnemiopsis* that invaded the Black Sea in the 1980s. There were hundreds, millions, so many they were uncountable.

That night a group of us layered on our heaviest wetsuits against the freezing-cold water of Puget Sound. We armed ourselves with underwater flashlights and slipped below the surface. When we found the bloom, it was as if the water grew thick, like bubble tea. Then we cut our lights and let our eyes adjust to the darkness. Soon I could see that every motion of my hands and feet ignited a flurry. Globs alight in glittering stripes spun away from me in all directions. I was in the middle of my own fantastic light show. I looked around in the glowing sea to share the magical moment with the ichthyologist surfer, but he was already far away. All I could see were the balls of bioluminescence careening off the back edges of his fins.

While he was floating in a rowboat near Friday Harbor, an idea jolted Osamu Shimomura awake. He'd used acid to solve the problem of ostracod bioluminescence. Maybe acid was the solution again. He returned to the lab. He acidified the squeezate and found that the glow stopped. When he raised the pH to neutral, it began to glow faintly. He was on to something. Cleaning up from the experiment, Shimomura dumped the neutralized squeezate in the sink. A blast of blue light erupted in the splash. There was seawater sloshing in the bottom of the sink. That was the secret. Shimomura later teased out the fact that calcium, which is abundant in seawater, was the magic ingredient that activated the jellyfish's glow.

Figuring out how calcium activated the glow required a lot of jellyfish. Each jellyfish contains only minuscule amounts of luciferin. To get the milligram needed for the job of characterizing the molecule, they needed two and half tons of jellyfish. Shimomura enlisted his whole family, including his wife, his son, and his five-year-old daughter, Sachi. They collected jellyfish from sunrise to sunset, over ten thousand in 1961 alone. "Our lab looked like a jellyfish factory," Shimomura said. Frank Johnson built a contraption that looked like a modified meat slicer to more quickly carve off the glowing edge of the jellyfish's bell. It "could cut ten times faster than by hand."

After five years of intensive jellyfish collection and squeezate study, Shimomura and Johnson discovered that the jellyfish luciferin is bound in a very stable complex with another protein they called aequorin. When a jellyfish is disturbed, pores in its cells open, allowing calcium from the seawater to rush inside. The calcium changes the shape of the luciferin-aequorin complex, releasing a blue photon that shouts "Help!" or "Come and get it!" depending who's receiving the call. In their excellent book on the history of biologically produced light, *Aglow in the Dark*, Vincent Pieribone and David Gruber describe the reaction: "The protein is like a gun with the hammer cocked; the calcium binding pulls the trigger." After the flash of light, the jellyfish cell pumps out calcium and pumps in oxygen, which resets the complex so that it's ready to fire again the next time a fish comes nipping at the jelly's tentacles.

Like many people in the United States, I've seen a television commercial showing elegant undulating moon jellies and claiming that these ancient animals hold the secret of memory loss. The key, the advertisement says, is glowing jellyfish protein that has been compounded into a dietary supplement. This protein is not harvested from jellyfish but is produced by bacteria containing jellyfish genes, and is called apoaequorin. While the company selling the apoaequorin-laden capsules provides no explanation of a mechanism, the protein's calcium-grabbing abilities may play a role. Calcium is important to our nervous systems, signaling the release of neurotransmitters. A few decades ago, some research suggested that as we age, we lose the ability to control calcium levels inside nerve cells. The product has undergone one clinical trial in humans, involving just over two hundred people who self-reported concerns about their memory. While the controls were somewhat weak and no one has tested whether the protein can pass through the blood-brain barrier, the company claims it's the top-selling brain-health supplement in pharmacies.

Although Shimomura had unlocked the mystery of the jellyfish glow, one big piece of the puzzle still remained. Shimomura's squeezate produced a blue flash. But in the wild, the jellyfish's glow is green. Shimomura surmised that something in the squeezate caused the blue light to shift to green light, a process called fluorescence.

While both fluorescence and bioluminescence glow, they are different processes. In bioluminescence, a chemical reaction converts chemical energy to light energy, forming photons where there were none before. In fluorescence, a molecule takes photons that already exist and changes their color. I think of it like a handoff. A fluorescent pigment takes the light of one color and passes it to another color. You've seen this at a black-light party, where colors seem to jump off fluorescent clothes. Pigments in the clothes take the black light, which has wavelengths too short for us to see, and pass it to a longer wavelength, such as day-glo green or neon pink, that we can see almost too well.

In the 1970s, Shimomura crystallized and characterized the chemical structure of the molecule that the jellyfish uses to transfer the blue light its luciferin makes to the green light that flashes. It is called green fluorescent protein, or GFP. He discovered that this GFP is a very special compound. Most fluorescent molecules are a chain of amino acids, with the part that shifts the color of light—called the chromophore—attached at the end. But in GFP, the chromophore is in the center of the chain. In three dimensions, the GFP is shaped like a soda can made out of 238 amino acids. The chromophore is tucked safely inside the can. In addition to its super-stable construction, this GFP is unlike other fluorescent proteins in that it automatically glows green in the presence of blue light. It doesn't need an enzyme. It doesn't need an additional energy source. It's a self-contained glowing unit.

Here the story shifts from Shimomura and the jellyfish to the microbial and molecular biologists who figured out how to install this self-contained glowing unit into the cells of other animals. Martin Chalfie at Columbia University worked on a millimeter-long worm, tracking the development of each of its organs and even some of its cells as the animal grew from a fertilized egg to an adult. The microscopic worm is made of just over a thousand cells and is transparent. A thousand cells aren't that many to keep track of, so scientists like Chalfie had learned a lot about how the animal's tissues grew and changed through its life. But there was a limit to what could be observed. Even through the best microscope, it was impossible to see proteins or DNA in a living animal. There was no way to tell when genes were being turned on and off.

One day, Chalfie went to a lunchtime lecture on a topic he knew nothing about: bioluminescence and fluorescence. When he heard about the self-contained green glowing jellyfish molecule, lights started going off in his head. He realized if he had the genetic code for a jellyfish's GFP, he could insert it right in front of a gene he was interested in following in the worm's chromosome. If the gene Chalfie wanted to follow was activated to form a protein, the GFP gene would be activated to form its protein, as well. Then, when Chalfie shined blue light on a cell that had the activated gene, that cell would fluoresce green. Because the worm was clear, the green glow would show through. It would be like inserting a new sentence or two at the top of an e-mail before forwarding it. When the e-mail was forwarded, the new message would go along with it.

A few years later, another scientist named Douglas Prasher sequenced the genetic code for GFP and sent some of the DNA to Chalfie. Chalfie inserted the jellyfish's GFP gene next to a section of the worm's DNA that codes for a protein that recognizes when the worm is being touched. When he shined a blue light on the worm, four neurons along the worm's body glowed bright green, showing exactly where the touch receptors were located. Chalfie could watch the worm grow and see exactly when the gene for touch was turned on and when it was turned off. It was a brilliant experiment. The worm's image was chosen for the cover of the journal *Science* on February 11, 1994. GFP was a new beacon in science. It allowed scientists to literally see how cells work.

Here the story gets even more colorful. Beginning in the early 1990s, Roger Tsien, a scientist at the University of San Diego, began tinkering with the GFP molecule, substituting amino acids along the length of its chain, as you might trade old charms for new ones on a charm bracelet. Some of his creations were failures; they didn't fold neatly into the soda-can shape, or the new protein didn't glow. But some of them worked. Over the years, Tsien created fluorescent proteins in all the colors of a fruit salad. He gave them names like mCherry, mStrawberry, mTangerine, mOrange, mBanana, mBlueberry, and mCitrine. (The *m* in front stands for monomer, meaning that the protein is just a single chain of amino acids.) This palette of fluorescence allows researchers to mark different proteins with different colors, so they can trace the activities of not just one gene, but multiple different genes as they are switched on and off in cells.

The uses for the fluorescent proteins are endless, and they have revolutionized biotechnology. Scientists have used them to study how bacteria divide, how flagella form, how cancer spreads, how Alzheimer's cells kill neurons, and how HIV infects cells. Scientists bred mice whose brains shine with different colors. Their technicolor nerves illuminate the complex tapestry of the brain in a "brainbow," which has helped us trace neurological pathways. Scientists have created bacteria that act like microscopic sniffer dogs, glowing in the presence of arsenic, of TNT, and of toxic heavy metals. They created glowing tests that identify devastating diseases like malaria and ebola.

Fluorescent proteins have also been used to create Frankensteins of all varieties. The local aquarium in Austin displays four axolotls, Mexican salamanders with a mane of ruffled gills. Two of them are genetically modified with jellyfish protein and glow green from shoulder to tail under the blue light that illuminates their tank. Zebrafish, originally engineered to turn fluorescent to help detect pollution in waterways, are now available for sale to light up your home aquarium. A swift Google search will show you images of glowing mice, rabbits, cats, pigs, and monkeys—all made possible by the glowing jellyfish gene.

The green-glowing proteins have even expanded their reach beyond living organisms in uses that are more akin to microchips or nanoparticles. Researchers are testing "tumor paints" that target cancer cells with fluorescence so that surgeons can see them more clearly and remove them with pinpoint accuracy. Iranian researchers are reported to be using green fluorescent proteins to build solar cells. And scientists in Scotland and Germany developed low-temperature, energy-efficient lasers using green fluorescent proteins.

Evolutionary biologists measure the success of a species by the number of offspring it produces, a measure called fitness. Michael Pollan, in his book *The Botany of Desire*, explained that one way species improve their fitness is by coaxing humans to help them proliferate. Domestication has increased the fitness of food plants, like wheat, corn, and potatoes. Herding has increased the fitness of cows and sheep. Since we like to eat them, we ensure these species' survival. Evolutionary biologist and rabble-rouser Richard Dawkins argued that it isn't the species that fights for fitness but rather the genes inside the organism that strive to make copies of themselves. While we usually think of domestication in terms of animals and plants, GFP has essentially become a domesticated gene. By being so useful to us, this single gene has circled the globe, transferred from Petri dish in one lab to cell culture in another to your neighborhood pet store. Beyond the physical locations where it's traveled, the jellyfish gene has invaded the tree of life, embedded itself in the genetic code of bacteria and yeasts, plants and animals. As a biotechnical tool, it has even attained an extra-corporeal existence. It's very possible that the jellyfish's green fluorescing gene is the most evolutionarily successful gene of all time.

In Stockholm in 2008, Osamu Shimomura stood tall and dignified, wearing a formal tuxedo and white tie, as he received his Nobel Prize in chemistry, along with Martin Chalfie and Roger Tsien, for the discovery, expression, and development of green fluorescent protein. A man whose youth was darkened by the worst devastation humans have ever wrought provided us with light far more powerful. The Nobel Committee compared the achievement to the discovery of the microscope by Anton van Leeuwenhoek in the seventeenth century. The green fluorescent protein literally shed new light on the vast and complex inner workings of the cell, processes that had been as invisible as the cells themselves were before the microscope opened our eyes to the miniature world. The Nobel Committee wrote, "When scientists develop methods to help them see things that were once invisible, research always takes a great leap forward."

Jellyfish Sense

Back in our dining room in Austin, it was feeding time for our new pet jellyfish. For once, the kids were interested in my world of jellies. Following the directions, we broke off a tiny bit of frozen jellyfish food and placed it in a dish. Drizzling a bit of seawater on the chunk melted it into a slurry of maroon baby brine shrimp. I thought about what a curious twist of evolution it is that jellyfish are more ancient than their food.

With a turkey baster, I showed Ben and Isy how to gently squeeze the food toward the underside of the jellyfish bells. I warned them to be careful not to get bubbles underneath, because they could rip the animals' delicate tissue.

"Look, you can tell they're eating," I said, pointing out the brick-colored bits that were accumulating like delicate piping along the oral arms.

Isy considered our three new pets. "How are we going to tell them apart?"

"The big one has one crooked oral arm," Ben said. "The middle-sized one is flat. The smallest is kind of round. And it's sort of upside down." I found that last observation worrying.

"We need to name them," Isy said. "We need a name for three things."

"Like the Three Stooges: Moe, Larry, and Shemp or Curly," I offered.

"Who?" Ben asked.

Eventually we decided to go with one of Isy's suggestions. The small one was Peanut; the fat one, Butter; and the one in the middle, Jelly. I'm sure we aren't the only jellyfish owners who have named their pets for a sandwich.

As I walked by the tank on my way from one side of the house to the other, I found myself stopping in my tracks, watching Peanut, Butter, and Jelly swim. It was relaxing and soothing, like having a miniature spa in my dining room. Soon I moved a lone chair so that it directly faced the tank. Sometimes during the day, while the kids were at school and I was supposed to be working, I sat watching the pulsing, the resting, the floating. What was life like for a jellyfish? What did they feel? Could they hear the sounds in the house? Did they taste the water? Did the jellyfish know I was watching them?

The answers existed along the edges of moon jellies' bells, which were scalloped like fancy lace parasols. If I looked very closely at the notches between the scallops, I could see a tiny intensification, a place where the bell looked a little less transparent. Under a magnifying glass, these spots are club-shaped structures called rhopalia (rhopalium in the singular), from the word for "bludgeon" in Greek. My moon jellies had eight, and while they tend to come in multiples of four, the number of rhopalia varies in different species.

Over the top of the moon jelly's rhopalia, there's a little flap, like a hood or an eyelid—rhopalia hold the gift of sight. We have just one type of eye, but ten different types of visual sensors exist in the animal kingdom. Jellyfish use at least three and probably four of them. My pet moon jellies had two types of eyes on each of their eight rhopalia—one shaped like a cup that looked up and a flat one that looked down—a total of sixteen visual sensors. However, for Peanut, Butter, and Jelly, vision was not high definition; it was a hazy thing. They probably couldn't see me as much more than a passing shadow.

Some jellies can see light without eyes. How they do it is mysterious. Scientists call it extra-ocular sensitivity, and I think of it as similar to the way we sometimes know that someone is watching us even when our eyes are closed. A jellyfish larva, a planula, doesn't have eyes or eyespots or any obvious visual receptor cells, but if you shine a light, it will swim toward or away from it, depending on the species. Even though jellyfish polyps don't have any apparent way to see light, they contract and wave their tentacles when light changes.

A lot of jellies make a daily vertical migration in the ocean in response to light. At night, they rise up to the surface to feed under the cover of darkness along with trillions of other planktonic organisms. At daybreak they sink back down into murky depths. Forget about the wildebeests or even the monarch butterflies. This vertical migration of the plankton is the world's largest migration. It occurs in all oceans twice a day. It's thought that for moon jellyfish—and probably other species as well—the cue to migrate is visual, because on cloudy days they remain near the surface, unaware that day has broken and that they should sink for the cover of darkness in the ocean's depths.

Changing light also makes jellies' eggs mature and their gonads swell. Some jellies spawn after a full moon. Many jellies, including moon jellies, respond to a sudden decrease in light by pulsing faster. Scientists call this the shadow response and think that it's a get-away-quick behavior to avoid becoming the lunch of a large light-blocking fish or turtle.

Although it's a murky world for most jellies, it's not quite as dim for box jellies. These hawks of the jellyfish have surprisingly complex visual systems, and they use them to hunt actively, much like fish. Box jellies are a particular class of jellyfish, about fifty species strong, that includes some of the most toxic creatures in the sea. Scientist Anders Garm from the University of Copenhagen has been studying the eyes of a Puerto Rican box jelly called *Tripedalia* for over a decade. Unlike its fearsome cousins, this Caribbean native's poison is fairly mild, which makes it a reasonable creature to study in the lab. "You have to kiss them practically to feel the sting. Then your lips go numb. Then you are fine," Anders told me.

In the journal *Nature*, Anders and his colleagues called the visual sensors of box jellies "a bizarre cluster of different eyes." It's a fair description. Most box jellies have six eyes on each of their four rhopalia, making a total of two dozen eyes. The six eyes are arranged in two rows of three, like the dots on a six domino. The four outlying eyes are fairly similar to the eyes of the moon jelly. They are cup-shaped, with pigments in the bottom of the cup that can absorb light. But the two eyes in the center column are surprisingly sophisticated, with roughly the same parts that make up the eyes you are using to read these words. These central eyes have a cornea, a lens, a retina, and an iris. What's more, if you shine a bright light on a box jellyfish, the iris closes down, protecting the eye from the excess light. The constriction is a little slower but not unlike what happens to our eyes when we walk into the sunlight. The top three eyes peer upward and the bottom three are angled downward, even if the jellyfish gets flipped completely upside down.

Perhaps the greatest technical achievement of the box jellyfish eye is its lens. If the clear proteins that make up the lens were packed in the simplest way—uniformly like balls in a gumball machine—then the lens would form a blurry image. No big deal; jellyfish nervous systems probably can't interpret a sharp image anyway. But the jellyfish lens doesn't produce a blurry image. Its proteins are packed more compactly in the center of the lens and more loosely around the outside in just the right manner so that all of the light rays that pass through the lens focus on a single spot. The box jellyfish lens creates an incredibly sharp, clear image. Anders and his coauthors wrote in *Nature*, "For such a minute eye, it is surprising to find well-corrected, aberration-free imaging, otherwise known only from the much larger eyes of vertebrates and cephalopods [octopus and squid]. The gradient in the upper eye lenses comes very close to the ideal [optical] solution."

What's even harder to understand than a jellyfish owning a perfect lens is this: If you map out where the rays of light go as they pass through the lens and deeper into the eye, the retina, which detects the image, is not positioned where it should be to make use of the perfectly focused image. It's too close to the lens. The optics are the same as trying to take a picture of your friend standing in front of a tree when the camera is focused on the tree. What you are trying to see is blurry. "The sharp focus of the lens is wasted by the inappropriate eye geometry," Anders wrote. Evolutionarily, perfect things don't usually happen by chance, but whatever it is that a perfect lens does for a box jellyfish is still a delightful and unsolved mystery.

Another mystery of the box jelly's complex eyes is how its simple nervous system processes the complex information it receives. Like a psychologist studying mice, Anders used mazes to try to find out. He set up a jellyfish obstacle course, poles of different colors and thicknesses at one end of a tank, and put the jellies at the other end. The flow of the tank pushed the jellies gently toward the obstacles. When the colored poles were in the tank, no matter the color or thickness, the jellies avoided them, never once bumping into them. When transparent poles were in place, the jellies bumped into them. He concluded that the jellies could see the colored poles and that their brains had enough computing power to steer around them. Using other obstacle courses, he discovered that the jellies avoided vertical and diagonal stripes but crashed into horizontal ones.

Anders also figured out that each type of jellyfish eye has its own specific purpose. The jellyfish actually use their top lensed eye to find landmarks for navigation. Those eyes stare out of the water at the trees on the shore to make sure the animal remains in the roots of the mangroves where their food lives. The lower lensed eye did all the avoidance work, keeping the jelly from crashing into roots. Two of the other four eyes are used as depth gauges, and the other two control muscles that make sure the jellies' eyes are always oriented correctly relative to gravity. A visual system built on "special-purpose eyes" was the way to square the complexity of the many box jellyfish eyes with their weak neurocircuitry.

"Humans have one set of eyes picking up enormous amounts of information, but then you need a complex nervous system to sort out the information afterward," Anders said. "The alternative is that you have eyes designed to pick up a certain type of information. You can filter the information very strongly in the periphery. So the information uptake is much less and much more specific. That's why you call it a special-purpose eye." Each special-purpose eye supports only "one, maybe two, behaviors. So if you need more behaviors, as most animals do, then you have to add more eyes." It's an entirely different mental architecture from the system we humans use, which requires a lot of brainpower to process information *after* it arrives. Instead, in jellyfish each eye is responsible for passing on only a certain type of message, allowing the lean neural circuitry to process complex messages.

A couple of years after we were married but before we had kids, Keith was out with a friend for dinner. I was at home, reading or watching TV. Suddenly I felt woozy, as though I'd had too much wine. The feeling came fast. By the time I called Keith and asked him to come home, the world was spinning madly. The only way to stop the feeling was to keep my eyes sealed shut and my head perfectly immobile so that the only inputs my brain received about my location were from the skin that rested on my bed. If I peeked upward or shifted my head a fraction of an inch, I would feel as if I were reeling and I'd puke. With my stomach empty and my eyes sealed shut, Keith managed to get me in the backseat of the car, where I could lie flat, and took off for the hospital. When he pulled up at the emergency room, I tried to look functional. I opened the door, hoisted myself out, and took one step. *Whomp!* I hit the curb like a domino falling over.

In school, I was taught about the "five senses"—sight, sound, taste, smell, and touch. But I don't remember learning about what might be the most important sense: proprioception, the ability to orient yourself in the world. People can live without sight or sound or taste or smell, but I don't believe it would be possible to live without the ability to know your orientation: what's up and what's down. One reason for a vertigo attack like the one I had is a malfunction in the inner ears. Behind our eardrum is a space where tiny hairs protrude inward from cells attached to nerves. The hairs are embedded in a layer of gel. Above the gel is a layer of crystals. When you move your head, the crystals shift, pulling on the gel, swaying the hairs, and sending a signal through a nerve. When the hairs bend left or right, you can tell, and you straighten up if you need to. When the hairs shift quickly, you know you're moving fast; when they're not moving, you know you aren't moving, either. Sometimes, the crystals detach from the gel layer, and collect in a nearby part of the inner ear that's also used in balance, the semicircular canals. When this happens, the crystals bang around in the canals and trigger signals that don't match the ones from your skin, eyes, and the remaining crystals that are still attached. Your brain reads "Does not compute" and sends you into a tailspin. A short-term version of this can happen when you spin too fast on an amusement-park ride. The crystals keep moving after the ride ends, making the world continue to circle even when you've stopped.

We share our sense of proprioception with an ancient organ found alongside the eyes on each of a jellyfish's rhopalia, a clump of cells known as a statocyst, which tells the animal its position in its three-dimensional world. Statocysts aren't unique to jellies; they are found throughout the animal kingdom, from worms and sea stars to clams and crabs, evidence that your orientation in the world is one of the most important things you can know. A statocyst is a hollow ball of cells, each with many little hair-like cilia pointed inward, toward the center of the ball. Around the outside of the ball, a nerve fiber attaches to each of the cells. Trapped inside the statocyst are a few grains of a mineral—in a jellyfish, it's gypsum—that are free to move. When a jelly is upright, the grains rest on the cells at the bottom, pressing down on the cilia there. These cilia send a signal to the nerve cells they are attached to, informing a ring of nerves around the jelly's bell that it is upright. If the jelly gets swept into a current and tipped sideways, the grains roll inside the statocyst, pressing on cilia on the side of the ball. Signals from these cells on the side instruct the muscles near the signal to paddle slower and muscles on the opposite side to paddle a little harder. The jelly rights itself. Its pulses return to normal when the nerve signal again comes from cilia at the bottom of the statocyst. The system is more complicated, of course, because jellies don't get information from just one statocyst on one rhopalium, but from statocysts on all the rhopalia around the edges of the bell. The jellyfish has to integrate the input from all those different signals in order to come up with a response.

Alongside the statocysts on the rhopalium, moon jellies also have a sensory spot called a touch plate, which is a field of cilia that reach out into the seawater. The cilia bend and move with the current that flows over the rhopalia, perhaps providing more information about position but also input about speed and turbulence. Perhaps they do even more. Sound is a movement of water, a wave that oscillates water molecules. Sound waves wiggle these cilia on the touch plate, just as they wiggle the hairs in our ears that help us perceive sound. Together, the touch plate and the statocyst form what might be considered a jellyfish's very primordial ear.

In late May 1991, meteorologists at Cape Canaveral, Florida, were monitoring three weather systems, any one of which could delay the launch of the space shuttle. At the same time, engineers were scrambling to fix a leak of liquid nitrogen from the propulsion system, which would also mean delays. Such holdups would mean trouble for Dorothy Spangenberg, a jellyfish scientist from Eastern Virginia Medical School. Dorothy and her team had prepared nearly 2,500 infant jellyfish and polyps for the mission. They were to be the first—cue the reverb—jellies in space.

While it might seem odd to send jellies into space, NASA had good reasons. Although no one in the West heard about it for years, the second man in space, Russian cosmonaut Gherman Titov, reported feeling a nausea similar to motion sickness. Later, U.S. astronauts reported similar symptoms: malaise, headaches, and vomiting during the first two or three days of their flights. This affliction became known as space sickness, and even today no one really understands what causes it, though microgravity is thought to be a culprit. The parts of the ear that provide our proprioception don't give the same signals when gravity isn't pulling with the same strength. Because it's not easy to get inside human ears, NASA tapped jellyfish as stunt doubles.

For Dorothy Spangenberg's microgravity experiment on jellyfish, timing was everything. Moon jellies develop their statocysts and touch plates about seventy-two hours after they are born. She planned to induce one group of polyps to form baby jellies three days before liftoff, one group just twenty-four hours before liftoff, and one group in the microgravity of space. The experiment needed to be timed to perfectly capture the different stages of development.

On the morning of June 5, launch day, the storms swerved. The skies were clear over Florida. The problem with the leak was solved. At 9:24, the spaceship *Columbia* roared into the sky. Astronaut Tammy Jernigan was tasked with rearing the baby jellies onboard. They were grown in two containers holding less than three quarts of water on the *Columbia*'s mid-deck. The containers sat inside an incubator about the size of a microwave oven, which held the temperature at a constant, comfortable 82 degrees Fahrenheit. Eight hours after liftoff, Jernigan topped off the flasks with iodine to induce the last group of polyps to strobilate into baby jellies. She recorded video of the infants as they pulsed about in their orbiting aquaria and then preserved some of them for further study back on Earth. Jernigan told me that in space the ephyrae moved differently from their cousins swimming in identical control studies back on Earth in Dorothy Spangenberg's Virginia lab.

On Earth, the ephyrae swam up to the tops of their test tubes, then relaxed and sank downward. In space, they swam in endless circles. The jellies didn't resume normal behavior after reentry, either. About a fifth of the space-born jellies swam strangely after they returned to Earth. They were uncoordinated; they spasmed; their arms were out of sync. Some could swim only in circles. Jellies that were hatched on Earth but grew up in space had fewer statoliths than their Earth-bred cousins. Growing up in space gave the jellies chronic vertigo.

It makes sense that jellies born in microgravity wouldn't develop the proper nerve and muscle connections to balance properly, but questions remained as to how and why and what we could learn about our own sense of balance from them. On a second jellies-in-space mission, astronauts observed two groups of polyps hatch into ephyrae without being induced to do so with iodine. On Earth, iodine causes the polyps to make a hormone, thyroxine. Dorothy Spangenberg speculated that thyroxine might play a role in the development of statocysts, but she never got the chance to find out whether her hypothesis was correct. Shrinking budgets scrapped future experiments.

When I sat in front of my pet jellyfish in my dining room, it was easy to become swept up in the elegance of their dance, the seemingly spontaneous motion that felt disconnected from my reality, movement driven by impulses so innate and alien as to be incomprehensible. But then I would focus on the rhopalia. It's a myth that jellyfish are faceless creatures. In some ways, their sensory experience is broader than our own. They sense their world with not just one but with many, many faces.

The Nerve of the Jellyfish

A couple of days after I set up my jellyfish tank, I sat in front of it accounting for the nearly imperceptible shapes of my pets like a mother hen. There was Butter, and there was Jelly. Where was Peanut? A glob rested near the bottom of the tank, hung up on the Kix-shaped pebbles that came with the system. I grabbed a chopstick from my stash of takeout disposables and gently prodded Peanut with the blunt end.

Peanut rose in the water but was very damaged. The side of its bell had a huge gash. Its oral arms hung askew. Its tentacles were shredded. Still, when I touched Peanut gently, it contracted feebly. Even damaged, its motion was somewhat coherent. Some food might help, I thought, might give it nutrition to heal. I dropped a bit of the red shrimp food below its bell, and soon saw an encouraging piping of prey appear along the edges that were still intact. I watched Peanut for a while, worried. Its lopsided swimming seemed to me courageous, though I also recognized it as a nervous response, like a doctor's tap on your knee to make it jump. What made the jellyfish contract at all? How was it moving?

Back in 1874, a young English biologist by the name of George John

Romanes had been spending a lot of time watching the contractions of damaged jellyfish bells himself. However, he didn't come to them through the purchase of a few moon jellies and a Kickstarter aquarium.

When he was just two years old, George's family inherited a large sum of money, propelling them from a modest Canadian lifestyle into one more similar to that of the Crawleys of *Downton Abbey*. His parents only sporadically sent George and his siblings to school, instead letting them discover what interested them in the museums, beaches, and fields of England and Europe. When he was about eighteen, someone finally decided to get George a regular tutor and discovered that he was very smart. George applied to Cambridge and was accepted. Though George started off with the intention of studying theology, he soon met a fellow student named Francis Darwin, who had a very famous dad named Charles. Francis seems to have convinced George to try biology instead.

Charles Darwin was about forty years older than George, but they shared similar intellectual interests, and a strong friendship grew between them. Early on, George was certainly starstruck by Charles and saw him as an intellectual mentor, perhaps in a way his freewheeling, distracted parents never were. For his part, Charles harbored an earnest fondness, almost a fatherly feeling, toward George. After Romanes was married, Darwin cautioned him against becoming a slacker, writing, "You have done too splendid work to turn idle," and enclosing in the letter five pounds and five shillings to buy some new equipment for his lab. I doubt that Romanes, wealthy as he was, needed the money, so it sounds a whole lot like a parental guilt trip to me—but a kind one.

I admit that I've tried to read Darwin's books, especially the canonical *On the Origin of Species*, no less than a dozen times in my life. It's been on and off my nightstand for years. I can't manage it. The sentences are thick and full, and my mind wanders away from the writing without digesting its meaning. In the correspondence between Darwin and Romanes, however, I discovered the brilliance, curiosity, and humanity of not only George Romanes but also Charles Darwin.

About halfway through university, George joined a newly formed neurophysiology lab at Cambridge and began studying the nature of nerves. At the time, no one knew much about the jellyfish nervous system, so George, who conveniently owned a summer home on a part of the Scottish coast where jellyfish often floated by, set himself the task of figuring it out. Darwin was curious about George's research on jellyfish nervous systems, in part because he was working on insect-eating plants, particularly the mechanism that causes a carnivorous plant's trap to spring shut, so a number of their letters discuss questions of jellyfish neurology.

Here's an example. George explains the behavior of a small jellyfish, just an inch and a half across, with four pinkish coils of gonads decorating its bell like garlands. This jelly has a cone-shaped mouth called a manubrium, which hangs off the underside of its bell like a chandelier. When George touched the edge of the bell near the tentacles with a pin, the jellyfish curved itself up in a "crouching attitude," contracting inward like a cat with its back arched so that its cone-shaped mouth was pulled inside its bell. Then the animal reached over with the tip of the cone and touched the exact spot on the edge of the bell that George had pricked. If George pricked a second spot, the tip of the mouth would leave the spot where the first prick was made and touch the next one with precise aim. George wrote to Charles Darwin, "It is interesting to observe when after such a series of irritations, the animal is left to itself, the manubrium will subsequently continue for a considerable time to visit first one and then another of the points which were irritated." If the idea of kissing your wounds comes to your mind, it does to mine, too.

George suggested that the reason for the repeated kisses is to sting or eat whatever was causing the pain, as the end of the cone-mouth is armed with many stinging cells. But how does the jellyfish "know" where it's been irritated? How does it remember where the pinprick was?

Romanes was a product of his times, a Victorian whose enormous curiosity and scientific conviction trumped sensitivity to animal welfare. He was a mad scientist when it came to his experiments on jellyfish, torturing them in every way imaginable: heating them, freezing them in a solid block of ice and then defrosting them, asphyxiating them, poking them with needles, and bathing them in chloroform, strychnine, nicotine, morphine, alcohol, curare, cyanide, acid, laughing gas, caffeine, and freshwater. George sliced jellyfish in half, in quarters, and in eighths. He cut out their gonads and the edges of their bells, turning them into a tube of tissue. He unfurled jellyfish bells like a gelatinous paper towel.

George wasn't oblivious of the way such treatment appeared, and spent several pages justifying his experiments in the introduction to his treatise on jellyfish neurology, *Jelly-Fish, Star-Fish and Sea-Urchins: Being a Research on Primitive Nervous Systems*, published in 1885. "I feel that it is desirable to touch before proceeding to give an account of my experiments, and this has reference to the vivisection [dissection of animals while alive] which many of these experiments have entailed. . . . The animals in question are so low in the scale of life, that to suppose them capable of conscious suffering would be in the highest degree unreasonable. Thus, for instance, they are considerably lower in the scale of organization than an oyster, and in none of the experiments which I have performed upon them has so much laceration of living tissue been entailed as that which is caused by opening an oyster and eating it alive, after due application of pepper and vinegar." I still don't feel great about George's torture of jellyfish, but I've eaten my share of oysters on the half shell, so let's proceed.

In what he called his "fundamental experiments," George shaved off bits of the jellyfish's edge to try to understand how the animal controls its motion. He discovered that if he cut off a thin edge of the bell all the way around, separating a ring containing the rhopalia from the rest of the bell, the ring made of the bell edge would continue to pulse, but the bell denuded of its edge would fall motionless to the bottom of the tank. George realized that he'd localized the key to the pulse to somewhere on the bell's outer edge. But that wasn't good enough for him. He began shaving off the rhopalia, one by one. Missing one rhopalium, the jellyfish could still swim normally. Missing two, three, four, five, and six, it still could swim. Even with just a single rhopalium left, the jellyfish could contract and move. But as soon as that last rhopalium was removed, the jellyfish fell motionless.

George concluded correctly that the trigger for jellyfish contraction lies in the rhopalia. Today, jellyfish scientists call the trigger a pacemaker, just as the signal from our own nerves that stimulates our hearts to pump with a regular beat (or an implanted medical device that does the same thing) is called a pacemaker. We aren't wrong when the pumping of a jellyfish stops us in our tracks. We are dependent on the same repetitive cadence to survive. Our hearts beat with its same primitive rhythm. My poor Peanut was pulsing because, although it was damaged, when I touched it with a chopstick, I stimulated one of the rhopalia. Even though the animal was wounded and dysfunctional, the signal from the pacemaker made its way to the muscles, causing a contraction.

George went on to discover that the wave of contraction caused by a jellyfish pacemaker is like a wave of enthusiastic fans in a circular sports stadium. A jellyfish's pacemaker starts contractions that proceed around the circle of the bell in opposite directions. When they meet on the far side, they cancel each other out. But only a damaged jellyfish has just one rhopalium; most have four or more, and each one sends its own wave signal to the muscles encircling the bell, telling them to contract. How does the jellyfish know which one to follow? When you look at a jellyfish's motion, it's not a hodgepodge of contractions. It's one elegant squeeze. It wasn't until the 1950s that scientists discovered the key to coordination among the pacemakers. The pacemaker that sends in the fastest signal overrides all the others and sets the pace for the entire jellyfish. If nothing unusual is going on, the rhopalia send a signal at a basal rate, and the jellyfish just pumps along at a relaxed pace, conserving its energy for more troubled times. If danger is sensed by the eyes, sensory touch plate, or statocyst on a particular rhopalium, the pacemaker nearest that rhopalium speeds up its signal, which overrides all the slower signals from the other rhopalia, and the jellyfish quickens its pumping and escapes.

There's something very democratic and modern about this multiple-pacemaker arrangement, evolutionarily ancient as it may be. It is a completely diffuse yet incredibly functional system. The jellyfish's nervous system is smart without being consolidated. No central brain is required to make a decision about what to do. Any pacemaker has the authority to speak up and take control if conditions demand it. That's an idea we're not used to, biased as we are to believe that intelligence originates from one centralized decision-making brain. But if there's anything that crowdsourcing and social media have taught us, it's that having eyes looking, ears listening, and brains thinking in many places and in various directions can be even more effective and elegant than a single authority. The collective wields great power.

In my final year of grad school, I snagged a lease on a rent-controlled studio apartment in a desirable zip code near the beach. It was still about two streets away from respectability, but it was much better housing than I'd had during my previous seven years in L.A. The walls were freshly painted, and the floor had new carpet. The kitchen and bathroom fixtures were cheap but clean. I felt mostly safe, especially if I sprinted past my drug-addled neighbors and didn't meet the gaze of the person who sometimes jumped out of the dumpster in the parking lot.

The ichthyologist surfer and I had finally fully broken up, but I still ached, especially when I ran into him in the hallways at school. To stave off the sadness, I called up old friends and begged them to keep me busy. Four of us started meeting regularly on Wednesday nights at the Coffee Bean & Tea Leaf on Santa Monica Boulevard. Over lattes, we discussed our shared condition of not having boyfriends.

"Enough of this victim stuff," my friend Stefanie said one night. "We need a plan." And she had one for us. Every Wednesday night, each of us was to arrive with a strategy for meeting someone datable in the following week—and not just anyone but someone we'd like to be in a relationship with. Our strategies were not to be decided alone. Each had to be approved by the group. And there was accountability: We were required to report on the results of our strategies from the previous week.

We shared our strategies around the table. "I'll agree to go on a blind date I've been avoiding." "I'll ask that guy Charlie I met last week to go rollerblading." "I'll go to a party I don't feel like going to." "I won't call my old boyfriend for the whole week."

We reported back. The blind date led to a second. Charlie was a fun rollerblading buddy, but was there more there? The party turned out to be a dud, except for this one whip-smart actor with a weird sense of humor. The old boyfriend had finally given up leaving unreturned messages.

In the weeks that followed—it being L.A.—I met and dated a string of sitcom writers. One wrote for *Caroline in the City*, another for *Mad About You*, and another for *Dharma and Greg*. They were all smart and sweet but had a lot more angst than I did, and something always went south. When one asked me why the San Andreas Fault couldn't be shored up with cement, the group agreed on a new strategy for me: no more sitcom writers.

After that, I went through a pretty long dry period, so I was given a standing strategy to say yes to anyone who asked me out, unless he was a sitcom writer. The strategy I proposed for the week I met Keith was to meet a friend who was visiting for work and had invited a whole gaggle of people to a bar in Hollywood. Maybe someone I might want to go out with would be there. Stefanie agreed to go with me.

In a big group of people, Keith and I found ourselves seated across from each other. He was funny and good-looking, with twinkling eyes and a sweet smile. He wasn't a sitcom writer. In fact, he had a job as different from either sitcom writer or ocean-science graduate student as you could get: He was a banker in downtown L.A. We fell into an easy conversation that had me laughing comfortably. But our fun was spoiled by an annoying friend Keith brought along, who made doltish comments about marine biology: "Do you study Flipper? Do you know Jacques Cousteau?" I mean, seriously.

When Keith asked if he could call me, Stefanie heard the question. I started to shake my head no, but her raised eyebrows said, *Strategy.*

"Yes," I answered, changing my life forever.

I may have been slow to respond to the nerves firing when I first met Keith, but back in 1875, George Romanes was purposeful about understanding the layout of the nerves within jellyfish tissue. And a warning—here's where George's mad-scientist mutilating tendencies really got going. He took a common moon jellyfish and sliced it eight times from the edge of the bell nearly to the center. Between those cuts, he sliced the jellyfish from the center outward nearly to the edge of the bell. His goal was to cut off any direct connections between nerves around the bell. But when he pinched the edge of the mangled tissue, he observed that "the contraction waves starting from the ganglion [his word for the pacemaker], continued to zigzag round and round the entire series of sections." Despite the severe lacerations, the jellyfish nerves could still send their signals.

George went on to slice and dice the jellyfish in truly sadistic ways. He even carved a jellyfish into the kind of spiral you get when you peel an orange in one piece. Even then, he reported, the jellyfish nerves were able to transfer their signals and contract muscle from one end to the other. You can nearly hear him exclaim across the decades, "Now in this experiment, when the spiral strip is only made about half an inch broad, it may be made more than a yard long before all the bell is used up in making the strip; and as nothing can well be imagined as more destructive of the continuity of a nerve-plexus than this spiral mode of section must be, we cannot but regard it as a very remarkable fact that the nerve-plexus should still continue to discharge its function." His observations that nerve signals could survive in even the smallest bit of jellyfish tissue led him to recognize that the jellyfish's nervous system was a very fine network of nerves. It had to be fine to pass a signal from one side of the jellyfish to another no matter how mutilated the bell was.

Later George learned to stain the nerves with gold so he could see them through a microscope. Describing what he saw, he wrote, "If the reader will imagine the first of the diagrams [of jellyfish] to be overspread with a disc of muslin, the fibres and mesh of which are finer than those of the finest and closest cobweb, and if he will imagine the mesh of those fibres to start from these marginal ganglia [the rhopalia], he will gain a tolerably correct idea of the lowest nervous system in the animal kingdom." We now recognize that the mesh, or nerve net, as scientists and textbooks call it, isn't completely evenly distributed. Nerves are more densely packed in a ring around the edge of the bell, where it acts as a circular central nervous system.

When Romanes looked very carefully at the stained jellyfish nerves, he saw that they were never directly in contact with one another. There was always a gap between nerve cells. Here, George made an extremely important inference about the way nerves function. He recognized that something had to be transferring the signal between the nerves, that "the physiological continuity is maintained by some such process of physiological induction." That physiological induction is done by neurotransmitters, and the gaps between nerves are now called synapses, but none of that was imagined in the 1880s, when Romanes was working on jellyfish. The term *synapse* wasn't coined until 1897, almost two decades later, and it is likely that his work was the foundation for that discovery. In 1932, another student from the same neurophysiology lab at Cambridge where George had studied, Charles Sherrington, won the Nobel Prize for his discovery of the synapse. Sixty years earlier, George had predicted that it must exist.

While working on his carved-up jellyfish, George discovered that if he damaged the jellyfish by applying a bit of pressure on the tissue at a certain point, he could stop the nerve signal. When he then irritated the pacemaker to send the contraction wave, it would reach that damaged tissue and stop. But if he continued to start a signal, at some point, a trickle of contraction would pass through the blockage. The next time, a bit more would pass through. Eventually the contraction would pass through the point that had been damaged as if nothing had ever happened. George concluded that the nerve fibers in the damaged region became more and more functional with effort, and in a fit of speculation he could hardly contain he said, "We probably have a physical explanation, which is as full and complete as ever can be, of the genesis of mind." He goes on to say that the first time intelligence dawned on the world, it had to have happened when some new line of nerves excavated their way through tissue. "Thus is it that a child learns its lessons by frequently repeating them; and thus it is that our knowledge is accumulated."

Today's understanding of how we create memories is not terribly different from what George described. When we have a thought, it is held in our short-term memory, where it can remain only while we are thinking it. For an idea to move into our long-term memory, we have to build new synapses, near-permanent connections between a series of nerve cells. The first time we have a thought and connect two nerve cells, neurotransmitters are passed from the tail of one neuron to the top of the next one, where there are only a few receptors to receive the neurotransmitters. The neurological signal is just a trickle. But catching a few neurotransmitters stimulates the receiving neuron to build a few more receptors and the producing neuron to make a few more neurotransmitters. The next time that synapse is used—the next time we have the same thought—more receptors are ready to catch the neurotransmitters and send the signal onward. The signal grows stronger. The more we think the same thought, the more neurotransmitters and receptors we produce at that synapse. And so it goes until the memory's physical structure is a robust synapse, connected by the flood of neurotransmitters and a field of receptors. At this point, the memory is recalled effortlessly, as if it has always been there. Yes, George Romanes was speculating wildly when he compared the neurological signal in a jellyfish to the accumulation of knowledge, but he was right.

A few hours after its feeding, Peanut was down on the rocks again. That night Peanut disappeared entirely. No more than 5 percent organic material, it left no corpse behind that I could see. A wisp of life, it evaporated into the tank.

I ordered a new Peanut, but when it arrived, it was not pint-size like Peanut. It was a beauty, with long, elegant oral arms and sweeping tentacles. I renamed the other two jellies to accommodate this new pet, which was clearly the Butter of the group. The old Butter became Jelly, and the old Jelly became Peanut. But my experiment as a jellyfish owner was ill-fated. Jelly-turned-Peanut shrank to a shell of its former self and then vanished completely.

For weeks I kept a watchful eye on Butter-turned-Jelly as its size diminished. I worried about sucking it into the vacuum hose when I cleaned the tank, and I enlisted the kids to keep a watchful eye as I swept the uneaten brine shrimp from the bottom. Even so, one day the little creature simply evaporated into the seawater like its comrades before it.

Butter was strong, big, beautiful. I was still fascinated by the way it gathered brick-colored food along the outer margin of its bell and down its four longer tentacles. It seemed almost magical when ten minutes later its four horseshoe-shaped stomachs were filled with the catch. I loved peering into its rhopalia and imagining that it was looking at my shadow, even hearing my voice. I imagined its pacemakers driving its contractions through the fine mesh of nerves.

Then one day I walked into the house and peeked into the tank, straining my eyes as always to find the clear creature in the water, looking for movement that signaled its location. Butter was inside out. Was the condition a sign of bad water? I started changing the water, slowly, cupful by cupful—a cup out, a cup in. Slowly Butter seemed to be pulling out of it. I checked every ten minutes. Butter wasn't getting worse. Within a couple of hours, it was normal again.

I redoubled my efforts to not only keep Butter alive, but make it even healthier. I cleaned the tank with every feeding, using a turkey baster to suck out the rotten bits of food that collect in the corners. Rather than once a week, I changed the tank's water every third day. At this point, Keith could see what I refused to. He said, "No more jellyfish. If this one doesn't make it, we aren't sending off to California for another jellyfish." I knew he was right. I also started to think that maybe a jellyfish didn't belong in a tank in my living room. There was something about the way Butter moved in the tank. It seemed trapped. It huddled in one corner often. Sometimes it just drifted at odd angles.

I wasn't the only one having trouble with the jellyfish tank. Online forums were reporting problems keeping jellyfish alive in the Jellyfish Art tank. A Facebook page was set up to discuss problems with the tank's design. A few years later, Alex sold Jellyfish Art to a company in Florida, which redesigned the tank entirely, flipping it on its side so that it's a vertical column. Other jellyfish tank companies have entered the market as well, with rounded-box designs and variations on filtration systems and pumps. Designing home aquaria that can keep jellyfish safe and healthy is a learning process. The industry hit an obstruction, but innovation continues to trickle through, not unlike the nerve signals in George's jellyfish.

One of the topics that George Romanes and Charles Darwin discussed frequently was heredity—the crucial missing link for Darwin in his theory of evolution. George often visited Charles at Down House, his home, south of London. They traded varieties of beets, potatoes, and onions and collaborated on experiments grafting different strains together to see what the hybrids would look like. Darwin knew that information was somehow transferred from parent to offspring. He also knew that variability among individuals was the raw material of natural selection. But he couldn't say how any of that happened. The mechanism of inheritance was a gaping hole in his theory, one that dogged him to the day he died.

The solution, the gene, was in fact known during Darwin's life. It was discovered by a Dutch monk, Gregor Mendel, and published in 1866, a full sixteen years before Darwin died. We know that Mendel even made forty copies of his meticulous work on pea plants, which proved the existence of the gene, and sent them to the most illustrious scientists in Europe. Darwin was probably one of those forty, although the manuscript was not found in Darwin's study after he died. It's possible that Darwin received it and then sent it off to another scientist; by the end of his life, he was a veritable lending library for scientific works, constantly receiving and mailing off manuscripts. Why wouldn't Darwin have read Mendel's treatise before sending it off to someone else? We don't know, but there are a few possible reasons. The manuscript was in German, and Darwin read German with difficulty. The text contained math, and Darwin was one of those biologists who didn't like math. So maybe he received Mendel's work, translated a bit of the German, saw the numbers, and thought, *Forget it*, letting the gem slip through his fingers. No one knows.

But fate fluttered the concept of the gene within Darwin's grasp one last time, the year before he died. Its messenger was George Romanes. George was working on an entry on hybridization for the *Encyclopædia Britannica* that covered work he and Darwin had performed on the topic. George asked Darwin for an up-to-date list of the scientists who had made significant contributions. "I think it is desirable to append a list of the more important works bearing upon the subject, and if I make such a list, I should not like to trust my own information, lest I should do unwitting injustice to some observing writers. If, therefore, you could, *without taking any special trouble*, jot down from memory the works you think most deserving of mention, I think it would be of benefit to the reading public."

Darwin wrote back, "I will send by to-day's post a large book by Focke, received a week or two ago, on Hybrids, and which I have not had time to look at, but which I see in Table of Contents includes a full history of subject and much else besides." Indeed, Wilhelm Olbers Focke's book did contain much else. It summarized Mendel's results on pages 108–111. But fate placed a fold of paper in the way. In the nineteenth century, books were printed on big pieces of paper, the size of a full sheet of newspaper. The pages were folded to book size and sewn together. When people read these books, they would use a penknife to slice open the folds, separating the pages. Pages 108–111 in Focke's book were never sliced apart. Romanes never read the report of Mendel's work, which described the unit of inheritance, the gene; otherwise he would have almost surely have shared it with his friend and mentor.

One April afternoon, the kids and I banged into the house after gymnastics and soccer and a bad meeting at work. Backpacks were flung around the kitchen. Notebooks littered the dining room table. The dog was fed. I turned on the oven, pulled some salad from the refrigerator, and hollered directions to anyone in my shouting range. I breezed past the jellyfish tank but didn't see Butter. No worries—I never saw Butter in its transparency until I looked hard.

Ben saw it first. "Butter's gone."

"What?!"

By the time Isy and I ran over, Ben had already discovered that the bubbler hose to the jellyfish tank had separated from the pump. We frantically scanned the water, looking in the corners, near the rocks, behind the bubbler. The tank was empty of life. Isy started to tear up. So did I. There was nothing left to do. I unplugged the bubbler, and the household hum quieted a notch.

Both George Romanes and Charles Darwin were students of theology as well as biology, and they both struggled to square evolution with Christianity. In one letter, Darwin asked George what he would say to a theologian who asked him to prove that matter and energy weren't originally created by God. After two pages of discussion, George concluded that there was no way to know, and he wrote, "But the more I think about the whole thing, the more I am convinced that you put it into a nutshell when you were here, and that there is about as much use in trying to illuminate the subject with the light of intellect as there would be in trying to illuminate the midnight sky with a candle. I intend, therefore, to drop it, and to take the advice of the poet, 'Believe it not, regret it not, but wait it out, O Man.'"

Life's Limits

"F*iumicino, per favore,*" I said to the cabdriver, asking him to take me to Rome's airport. I savored the Italian words on my tongue like a last spoonful of hazelnut gelato. No more would I walk to work through winding cobbled streets, across the Tiber on a centuries-old bridge, past the graceful Colosseum on the way to the United Nations' Food and Agricultural Organization, where I had come to work with my PhD advisor while he was on sabbatical. In a couple of days, my commute would be passing by another coliseum, the boxy Los Angeles sports arena, during a grinding stop-and-go along the 10 Freeway.

At the airline check-in counter, I was the only one in line. "The entire airport is on strike," the attendant told me. "Your flight is delayed." Italy comes with its delights but its problems too. Planned strikes are one of them. I window-shopped until my ugly overstuffed computer bag started digging too angrily into my shoulder and forced me to sit down. I had bought my first laptop for this trip, a heavy early-'90s model, and I was hauling around too much: printouts of computer models I'd run, journal articles, and plastic computer diskettes—now, thankfully, extinct. Hours passed. I bought chips from a machine for lunch, crunching back a fear about this flight. A week earlier, TWA Flight 800 from JFK bound for Rome had exploded and crashed into the Atlantic Ocean shortly after takeoff. The cause was still unknown. My TWA flight was going in the opposite direction, from Rome bound for JFK. The shared airline and the reversed route, and the ongoing mystery of the crash, had put me on edge.

Around dinnertime, according to some prenegotiated labor-union arrangement, the strike ended. We lined up and boarded the plane. I shoved my bulky computer bag under the seat and tucked my novel into the seat-back pocket. An Italian woman, Daniella, a little younger than I, with long dark hair, sat in the aisle seat. She said she was going to the United States for the first time. The pilot reminded us to relax and enjoy the flight. I pushed down my fears as we took off. Soon smells of international food service drifted down the aisle, predicting pasta. I was starved for real food after grazing from vending machines all day. When the flight attendant appeared at my row, I reached hungrily for the plastic tray.

A suite of dings sounded. She looked up, tray hovering.

More dings. I looked at the face of the flight attendant. Not assurance. Not nonchalance. "Flight attendants, please return to service areas." She pulled away the plastic tray. It didn't matter. My appetite disappeared.

"Flight attendants, please prepare the passengers for an emergency landing."

Daniella and I grabbed hands, instantly no longer strangers. A man behind us yelled, "There's a bomb on the plane!" A group of nuns returning from a trip to Vatican City seated in the row in front of me feverishly worked their rosaries. I wished I knew how to pray.

Now an announcement over the speaker in Italian. Daniella translated, "We will be landing in Paris, Orly. In fifteen minutes."

Fifteen minutes. I was not ready to crash land. More than that, I was not ready to die. And from a deep place in my body—not my mind—two thoughts surfaced about what I would leave behind if I died: My half-written dissertation on my computer didn't mean as much as I'd thought. If the printouts of models stuffed in my computer bag crashed with me and were destroyed, the world wouldn't change. And then, a thought that had never formed so perfectly in my mind before: *I needed to be a mother.* I was single at the time, broken up with the ichthyologist surfer but hadn't yet met Keith. I couldn't die without having a child.

The full moon was framed by the plastic edges of my porthole. Its intense glow was an untouchable, uncontrollable sign of the reality that existed outside this blighted aircraft. The platinum orb was as close to God as anything I have ever known. *Please,* I prayed to the Moon. *Please let me live. I am not done yet. I have more to do.* The plane banked then, stealing the Moon from my view.

The flight attendants shouted through the rows. "Heads down. Heads down. Get into the crash position. Now!" With my face smashed into my knees, I braced my feet on the floor. I was aware of the pressure of the seat against my thighs and the air on the back of my neck. I felt my stomach in knots, the bite of bile in my throat, my sweaty palm clenching Daniella's. I heard landing gear whoosh open, and within seconds the plane collapsed onto the ground. There was an explosion, and another. Daniella and I screamed. The man behind us bellowed, "It's a bomb!" And then everything went dark.

With my head still pressed against my knees, I peeked up at the window. The reason for the darkness was a coating of flame-retardant foam. Slowly heads popped up in seats around us as we reached the collective realization that we were still alive. But no cheer rose through the crowd. Rather, feelings of confusion, relief, and exhaustion washed through me, through the plane. Newly bonded in the experience of survival, we speculated quietly. Rumors spread: bombs, fire, terrorism, near–midair crash with another plane. As for the nuns, the entire row was fast asleep, worn out from praying with all their hearts.

In the end, the problem was one of instrumentation. The pilot received signals that there was a fire on board. There wasn't, but he landed because he couldn't be certain otherwise. It was the right decision, but the flight scarred me. For a decade afterward, I popped a Valium to travel by air. Even now, I usually drink a glass of wine before takeoff. But in retrospect, the disaster was a gift. I had the chance to reckon with what it would mean to die. I had the chance to take stock of what mattered in the face of my own mortality. I had the chance to really consider what I wanted to leave behind. And then I was given the chance to act on what I saw. Yes, it was an emergency, but it also was a landing.

A few years into my jellyfish research, Keith had a business trip that took him to Rome. I connived to turn his business travel into another jelly research opportunity for me. We cashed in some frequent-flier miles and farmed the kids out to my sister again. As we circled over the pinot grigio–colored fields near Fiumicino, I felt a familiar panic rise in my chest. I gripped his hand on one side and the armrest on the other. My palms were dripping with sweat. I planted my feet firmly on the floor of the plane.

"You going to hold us up in the air with that vise grip?" Keith asked. His humor was soothing. When we first started dating, Keith had suffered through my airplane panic attacks, his fingers in the clutch of my soggy hand. It had been a while since I'd been so anxious, though.

"It's returning to where it happened, I guess." I closed my eyes, forced myself to breathe slowly. We landed with a little bump. Keith shook out his cramped hand, then wiped my own sweat on my sleeve.

"Gross," I said, feeling relief.

That my sense of mortality came flooding back as I landed in Italy had a certain relevance, because I had come here to meet two scientists who study a jellyfish that can technically live forever. While Keith went to his business meetings, I was going to travel down to the heel of Italy's boot to the University of Salento in Lecce. There I would meet Stefano Piraino and Ferdinando Boero, who study a tiny jellyfish the size of the tip of a ballpoint pen, one that's nicknamed the immortal jellyfish.

Trim and silver-haired, Stefano walked into the lobby of the hotel where I was staying and introduced himself. He gestured to his ride outside, and I saw that it didn't have any doors. Or windows or roof, for that matter. I took the shiny black helmet he handed me and fastened it on. Yes, I did feel nervous climbing onto the back of a motorcycle behind some Italian guy I'd just met. I threw my leg over the seat, grateful that I'd decided to wear black jeans and not a skirt. I wasn't sure if I should grab his shoulders or what. As we cut into traffic, I gripped the bottom edge of my seat. Then Stefano accelerated. This was no *motorino.* The hell with it; I held on to his shoulders, my mortality in the forefront of my mind for the second time in just as many days. We slipped through the twisty streets of Lecce, passing lines of cars, tilting into turns. I hadn't expected to go so fast. I hadn't expected to like it so much.

The story of the immortal jellyfish starts in 1988, when a marine biology student collected a minuscule bell-shaped medusa with a smattering of thin tentacles and a pinkish chandelier of gonads from shallow water near Genoa on Italy's northwest coast. One Friday, he left the medusa in a bowl of seawater, forgetting to put it back in the refrigerator for the weekend. When he returned on Monday, the medusa was missing. But it hadn't completely disappeared like my pet jellyfish. This medusa had been a hydromedusa, which makes a slightly different sort of polyp—lankier, with wormlike roots called stolons. The bowl contained a stolon.

That was odd. Jellyfish are known to cycle through life in order: a fertilized egg grows into a furry Tic Tac larva, which metamorphoses into a polyp, which buds into swimming medusae, which produce eggs or sperm and then die. But there hadn't been enough time for the medusa in the bowl of seawater to spawn, grow into a larva, and end up a polyp over the weekend. Those transformations take weeks. In order for a polyp to end up in the bowl of seawater, the jellyfish had to have reverse-aged, like Benjamin Button, morphing backward through its life cycle from medusa to polyp.

It has been known for centuries that jellyfish don't always color inside the lines when it comes to their life cycles. Some species skip the polyp stage, going straight from planula to medusa. Many skip the medusa stage, remaining a polyp through old age. Polyps can bud from other polyps. Medusae can bud from the underbellies of other medusae. In 1909, German scientists reported that jellyfish ephyra, similar to the kind I saw popping to life at Monterey Bay Aquarium, can age backward to polyps. Despite all the plasticity in their life cycle, scientists had believed that there was a limit, that once a medusa reached reproductive age, those sorts of unusual transformations would be impossible. Once an animal became mature enough to produce eggs or sperm, it was thought, the only option was to spawn and die—that is, until the jellyfish left sitting on the counter for the weekend rejuvenated itself.

Stefano discovered a colony of the same jellyfish near his lab in Lecce—and as with Ponce de León's fountain of youth, its exact location remains a bit of a mystery to the general public. But Stefano knows where it is, and it has allowed him to very carefully study the life cycle of the animal. In the lab, Stefano and his collaborators watched the jellyfish morph from polyp to medusa and back to polyp and back to medusa and back to polyp, without ever going through the spawn-and-die part of their lives. They wrote, "This process would be hardly more remarkable if a butterfly were able to revert to its caterpillar stage. It must be considered a true metamorphosis, but in the opposite direction to larval metamorphosis." For all intents and purposes, these jellyfish were immortal. Still, Stefano said that he always cautions people that *Turritopsis* can be killed and do die, by infection or predation, among other possibilities. "If they were truly immortal, the ocean would be completely full of *Turritopsis*, and we don't see that." But at least theoretically, the jellyfish can morph forward and backward through their life cycle forever. One scientist in Japan has had the same *Turritopsis* in culture in his lab for dozens of years.

Recently, we learned that this proclivity for agelessness might not be constrained to just one species of small jellyfish. In 2016, a Chinese graduate student, Jinru He, neglected a moon jellyfish medusa. After a couple of days, the medusa sank to the floor of the tank and stopped moving. The animal broke into pieces, and any normal person would have considered it dead and washed it down the drain. But not this scientist; he kept watching. After a couple of months, the detritus of the medusa began to reconstitute itself like a phoenix rising from its ashes. Tentacles emerged. A mouth formed. Eventually perfectly healthy polyps sprang to life from the medusa carcass. Rather than simply die, the jellyfish seemingly reversed its life cycle, going from the degraded medusa backward to the polyp stage.

Like animals, individual cells proceed through a life cycle. All types of cells are born as generic *stem cells*, like a lump of dough with a lot of potential. Specific genes turn on and off inside each stem cell, changing it into a muscle cell, a skin cell, or a nerve cell in the same way a baker molds the dough into a pizza crust, a loaf of bread, or a pretzel. Certain steps are required for the transformation of dough, be it spinning it in the air, rolling it into a snake and giving it a twist, or mounding it into a loaf pan. And just as when a pretzel is removed piping hot from the oven, once a cell reaches its final configuration, it's fully baked. You can't turn a pretzel into a pizza any longer. Likewise, a muscle cell can't morph into a nerve cell.

Stefano wanted to understand what happens to the cells inside the jellyfish's body when it goes through its reverse aging. Was there a limit to the life of a cell for *Turritopsis*? Are *Turritopsis* cells ever fully baked? The answer—incredibly—seems to be no. Normally, the genetic switches that control the transformation of an embryo into a larva or a larva into a polyp are switched on in an order that is irreversible, Stefano explained. "But *Turritopsis* cells can hit the rewind button." When the *Turritopsis* medusa rejuvenates, muscle cells, for example, turn certain genes on or off, essentially unbaking the cells and reverting them to doughlike stem cells. Then those stem cells re-form into new and different cells in the polyp.

The idea that fully cooked cells can become stem cells again has enormous and tantalizing possibilities for medical research. If we could hit the rewind button in our cells, we could open the door to all kinds of cures for diseases in which cells have gone awry, ailments like Parkinson's disease and cancer. "Cancer is cell proliferation without rules," Stefano said. "It is uncontrolled growth. There is no plan for what to do with the new cells." But in *Turritopsis*, the reprogramming of one cell into another kind of cell "is part of a controlled pathway." We just need to learn what those controls are.

While I was in Italy, Stefano mentioned that one of his collaborators on *Turritopsis* had just moved to the Texas coast and was working outside Galveston at Texas A&M University's marine lab. That collaborator was Maria Pia Miglietta, who in 2009 discovered that the *Turritopsis* living in the waters off Japan, Panama, Florida, Spain, and Italy were all nearly identical genetically and classified them as the same species. So while the oceans are not filled with the immortal jellyfish, she confirmed that they have spread worldwide. Maria Pia and her coauthor wrote that the conduit for the spread was us, ships' ballast water, just as it was for the spread of the comb jelly *Mnemiopsis* from the Atlantic to the Black Sea and beyond. Unlike the comb jellyfish, the immortal jellyfish posed very little threat to the ecosystem, because it is so small—or to us, because its sting isn't painful. However, the immortal life cycle, the ability to flip back and forth between a medusa and a polyp, probably allowed it to withstand the stressful conditions in a boat's hull and likely helped the jellyfish achieve its global range.

Back in Texas about a year after I met with Stefano, I drove down to Galveston to meet Maria Pia. I wanted to know how the work on understanding the genetic switches that made the jellyfish immortal was progressing. She greeted me in the Marine Sciences building, a modern facility with beautiful mosaics of underwater scenes decorating the walls. About my size, but with a mane of curly dark hair, Maria Pia led me up the staircase that spiraled up the middle of the building to her office.

She told me that her work on *Turritopsis* genes was still in progress, and was a little hesitant to tell me specific details. She didn't want to say anything that might be contradicted once the data had been fully crunched. But she did tell me the direction the research was headed.

The plan she's following originated with two Japanese biologists, Shinya Yamanaka and his student Kazutoshi Takahashi. In the early 2000s, they injected mouse skin cells with between four and seven pieces of protein called transcription factors. Transcription factors attach on to DNA and control which genes are flipped on. The Japanese scientists discovered that the presence of just those few proteins had the power to transform skin cells backward—to the lump-of-dough stem-cell stage. And from those stem cells, scientists were able to grow nerve cells, blood cells, and heart muscle cells. For this rule-shattering work, Shinya Yamanaka received the Nobel Prize in 2012.

One major roadblock, Maria Pia told me, is that most of the work on transcription factors has been done on cells grown in a culture dish. In the real world, animal cells exist in the ecosystem of bodies. They are in constant conversation with the nearby cells, exchanging molecular bits of information that tell the cell how to function, much the same way the information that comes in through our eyes, ears, nose, mouth, and skin tells us how to function. "To really understand how one cell becomes totipotent [a stem cell] and then becomes something else, you need to know how it integrates in an organism," Maria Pia said. Yamanaka and a collaborator pointed out that same problem in a recent paper. They said if we really want to understand the fates of cells, we'd be wise to study an organism that already has a mechanism for creating stem cells from mature cells: *Turritopsis*.

So that's what Maria Pia and Stefano are doing. They have collected medusae and polyps of *Turritopsis* and are looking at which genes are active in de-aging, paying particular attention to the pieces of protein Yamanaka identified that have the power to turn mature mouse skin cells back into stem cells. Maria Pia listed the questions they hope to answer. "We want to understand the role of the [Yamanaka] genes in *Turritopsis.* Are they there? Are they not? And if there are only two or three genes, what is the role of each of these factors? And can we use *Turritopsis* as a model system to understand the behavior of the genes? That's where we are going." But until the data are crunched and the experiments are replicated, the secret to immortality remains the mystery it has always been.

While Stefano Piraino is soft-spoken and technical, his collaborator on the immortal jellyfish at the University of Salento is nearly the opposite. With a spiky white beard, close-cropped white hair, and ruddy cheeks, Ferdinando Boero, who goes by Nando, is as animated as he is clever. His speech is peppered liberally with "So c'mon," hand raised in the air, eyes toward the sky. Nando and Stefano worked together on the study on *Turritopsis*, showing how the jellyfish reverses its life cycle, but the day I met him, our conversation wandered beyond the molecular limits of life demonstrated by the immortal jellyfish into the vast seas of jellyfish science that suggest that humans are bumping up against biological barriers on the scale of ecosystems.

When I began researching jellyfish, Nando's articles jumped out at me and at times even made me laugh out loud. One of his recent titles said that "time is an affliction," and the paper further explained that "information isn't knowledge; knowledge isn't wisdom." These phrases sound like song lyrics because they *are* song lyrics—written by musical genius and counterculture icon Frank Zappa. To say Nando is a Frank Zappa fan is something of an understatement. Nando sees Zappa as a philosophical guide, an intellectual icon, and maybe even a god.

After graduating with his PhD in the classification and naming of hydrozoan jellyfish, Nando signed himself up for a research trip to California with the express purpose of meeting Frank Zappa. He had a plan to do it too. He would discover a new species of jellyfish, name it after Zappa, and use the honorific as a calling card to meet his idol. And that's exactly what happened. A short time after arriving in Bodega Bay in Northern California, Nando discovered a new species of jellyfish with a tiny clear domed bell about a quarter of an inch wide. Its tentacles hang down in a delicate fringe. The polyp from which the medusa emerges is long, thin, and elegant. Nando classified the animal into the genus *Phialella*, and then dubbed the species *zappai*. He sent a letter about the jellyfish to the Zappa residence in Laurel Canyon outside of Los Angeles. And according to plan, an invitation to visit soon followed. Out of this most idiosyncratic meeting, a real camaraderie was born, and the scientist and the musician remained good friends until Zappa died in 1993.

Toward the end of his touring career, Zappa performed in Genoa, Nando's hometown, so Nando made sure he was in the audience. Before the show, Zappa announced that he would dedicate the concert to his marine-biologist friend. Then he riffed on a song he'd originally called "Lonesome Cowboy Burt," retitling it "Lonesome Cowboy Nando" and switching up the verses to include references to marine biology, jellyfish, and Nando's lofty standards for judging the quality of pizza.

It's no wonder Zappa and Nando got along so well. Like Zappa, Nando pushes boundaries in a way that's both smart and skeptical. Like Zappa, Nando is brash and funny but committed and serious. His mind teems with theories, not just on immortal jellyfish, but also on the role of humanity as perceived through the lenses of biology and philosophy. His unusual Holy Trinity begins with Frank Zappa, of course, but it also includes Charles Darwin and George Carlin. His devils include the fields of physics and mathematics, which I guess makes me something of a Satan. He's not much of a fan of economics either.

"You know what economists call natural resources?" Nando asked after we'd been talking a while.

I didn't.

He answered his own question. "They call it an 'externality' and they write it off. It has no value. And this is a lie. And we let them get away with it." Several years ago, Nando struck up a conversation with an economist on a train. The economist said that in accounting, natural resources—commodities like seafood or coal—have no value. Economists may value the effort required to obtain a resource, but until they've captured it, it has no value at all. Zero. Nando isn't the only one to recognize this economic construct. Earlier in the year, I attended a conference called South by Southwest Eco, which seeks to bring together people with creative solutions for the environment. The famous ocean conservationist Sylvia Earle spoke on a panel about aquaculture and fisheries. She didn't use the word *externality*, but she brought up this same point. Fish in the ocean have no inherent value, no monetary value, until they are caught.

This idea runs counter to what ecologists understand about the way organisms work in the environment. Fish and shrimp and kelp—things we regularly catch and eat—have a real value in the ocean. Kelp provides food and shelter for shrimp and fish. Fish keep populations of shrimp and kelp in check. Shrimp are food for fish and, like fish, when they die they are decomposed into fertilizer for the kelp. These connections are what make a functioning ocean. Without a functioning ocean, fisheries can't exist. So how can natural resources be worthless?

In economic theory, externalities are defined as "the cost or benefit that affects a party who did not choose to incur that cost or benefit." In textbooks, these spillover effects are usually things like the air pollution created by factories or the medical care required by cigarette smokers. With regard to fishing, the party who did not choose to incur the cost of the loss of fish is everyone who depends on the ocean. So, everyone.

Nando has called out the economists on the fundamental problem with the idea of externalities. He wrote a book, in Italian, whose title may be translated *Economy Without Nature: The Great Swindle*. That title pays homage to Darwin, and the book describes the ways in which the economics of capitalist systems, which favor unbridled growth, opposes natural systems. When Darwin's *On the Origin of Species* was published in 1859, the word *ecology* didn't yet exist. Ernst Haeckel coined the term years later. So Darwin called the interactions between animals and their environment the "economy of nature." In his book, Nando wrote of the incompatibility between today's economic systems and the biological limits of our planet:

> Economy and ecology, then, despite having the same root, are antithetical, are against each other. Ecology, on the one hand, embraces nature. . . . Economics, on the other hand, takes away from nature. It doesn't observe the constraints and limits; it outsources and pretends that externalities—in other words, nature—does not exist. . . . Generation after generation, we have improved our way out of nature, domesticating plants and animals. We have replaced nature with the few species that we need. Slowly we nibbled it away, continuing to multiply. Go forth and multiply is in the nature of all species. But there are limits. Perhaps we have reached that limit.

Nando's recognition of the incompatibilities between economics and nature started when he was working on his degree in marine biology at the University of Genoa. He studied the data collected from the cooperative of tuna fishermen who worked out of a shed on the city's docks. The fishermen's bookkeeper kept detailed records of the catch between 1950 and 1975, noting every fish caught every day. Some years the catch was great, as in 1955, and some years were bad, like 1963, but if you run your eye across the data as a whole, there's no trend. However, if you break the catch into different groups of fish, you find something very different. The top predators, big tuna and sharks, decreased by 80 percent. By 1974, there were no big sharks left in the catch at all. Released from pressure from their predators, the smallest catch, like anchovies, increased fourfold and squid tenfold by weight. Because the smaller animals that replaced the top predators reproduce more quickly and have shorter lives, they are naturally more variable. This erodes the stability of the entire ecosystem.

What removed the top predators and destabilized the ecosystem? In the early twentieth century, fishing shifted from individual- and family-owned fishing boats, and local co-ops like the one that ran Nando's tuna trap, to corporation-owned businesses. New fishing tools were inadvertently invented during World War II. Radar, developed by the British Royal Navy, was used to help fishermen navigate. Sonar, first used in submarine warfare, allowed fishermen to find schools of fish deep below the surface. Plastics, synthetic rubber, and nylon twine made mending nets easier and the cost of new nets cheaper. The ability to freeze food was crucial, allowing fishermen to venture farther and return with a catch that wasn't spoiled. In the 1960s, the Soviet Union began building football field–size vessels that could remain at sea for half a year, freezing their catch in giant holds. Fleets of smaller boats serviced these tankers, supplying provisions and taking away frozen catch. These floating fish-processing factories traveled anywhere in the world, reaping fish from the oceans with nets half a mile across.

On this topic, Nando waxed biblical. "When God put Adam and Eve in the Garden of Eden, he gave Adam one job: to name everything. You could say, God asked Adam to be the first naturalist. But God put limits on Adam. He said, 'You can have anything in this garden but the fruit that grows on that tree.' Well, of course Eve ate the fruit. She went over the limit. And the punishment was labor. Adam and Eve were kicked out of the garden. Instead of being gatherers like they had been in the garden, they became farmers. And we continue this work of agriculture on the land until today."

Nando continued: "In the ocean, we still have the Garden of Eden. We are still able to hunt and gather from the sea. The problem is that the same rules apply. There should be a limit in the sea like on land, but we don't see this limit. Industrialized fishing is pushing us toward the limit, and we are about to exceed it—to be kicked out of this second paradise.

"We are exceeding our boundaries in other ways too," Nando said. "Our metabolism is based on combustion. We cannot avoid our corporeal combustion. But we also produce extra-corporeal combustion. It started with the domestication of fire, but now it has expanded to the form of burning fossil fuels. We have to move beyond this age of combustion. We have to stop our extra-corporeal combustion. The jellyfish are an alarm that says we are mismanaging our ecosystem. They are a symptom. The cause is our stupid behavior. And there is nothing else."

By this point, we had been talking for hours. It had been a long day of immortal jellyfish, fisheries economics, and philosophy. My head was tired, and I needed time to digest his ideas and find the gems among the liturgy, most of which was depressing. Nando sighed. He lives in this world every day; a man whose thoughts are strange and strong and deep needs respite too. Laughter is a good place to find it.

"You know George Carlin?" Nando asked.

"Sure."

He tapped his fingers on the keyboard of his computer, pulling up a YouTube video. "This is the very nice one, the one about saving the planet."

Together, we watched a gray-ponytailed, black-clad, laser-eyed George Carlin on a New York stage in 1992. He preached his dark comedy into the microphone. In this act, he was exasperated with humans who worry about nature, about our arrogance: "Haven't we done enough?" He railed against environmentalists' pretentiousness, about people—yeah, I'm one of them—who care about the whales and snails and bees and trees. Given the mighty forces that drive our planet—volcanism, plate tectonics, magnetic storms—Carlin said our penchant for focusing on small issues was infuriating. "The planet isn't going anywhere." He stared out of the computer with that fierce stare of his, confronting the inescapable mortality of our species. "We are. Pack your bags."

But Carlin concluded with a hopeful thought, because no one, not even a truth teller like Carlin nor a realist like Nando, can leave the stage on such a cynical note. "I don't worry about the little things: bees trees whales snails. I think we are part of a greater wisdom than we will ever understand. A higher order, call it what you want. Know what I call it? The big electron. It doesn't punish. It doesn't reward. It doesn't judge. It just is. And so are we. For a little while."

Nando was laughing out loud, gesturing at the screen. When the video ended, he leaned back in his chair and said, "I fully concur with George Carlin. He's as great as Frank Zappa." He raised his hand and eyes toward the heavens: "So, c'mon."

Strobila

The Bottom of the Wave

The National Center for Ecological Analysis and Synthesis, also known as NCEAS, was housed in a hip old building in the middle of the high-end shopping district in downtown Santa Barbara, California. To me, it was an unlikely spot for a place that specializes in crunching numbers, but there it was, with a Macy's on one side, a Banana Republic on the other, and a Tiffany's around the corner. When I first met with Monty Graham in Alabama, he had told me about the project with NCEAS to determine whether global jellyfish populations were on the rise. This workshop was one element of that project, an attempt to get a grip on their ecology. Rob Condon, whom I had met over a fried-clam lunch in Alabama, had invited me to this workshop, which focused on jellyfish ecology. It would give me a chance to dig deeper into some of the biggest mysteries about jellyfish: Just what do they *do* out there in the ocean? What role or roles do they play in the ecosystem?

The elevator to the NCEAS offices was beautiful vintage copper, embellished with an Art Deco motif that reminded me of Ernst Haeckel's jellyfish drawings. I pushed the call button, and the doors slid smoothly open. Peeking my head around the door of the conference room where the workshop was getting started, I recognized most of the members of the group. Rob Condon greeted me with a generous Australian-accented "Hullo!" PhD student Kelly Robinson was there too. I'd met her in Alabama when she was getting started on assembling the world's largest jellyfish database. A scientist whom I hadn't seen since grad school jumped up with a shout of recognition when I entered the room. Her name was Tammi Richardson, a phytoplankton biologist I'd worked with during my phytoplankton days. Her interests were spilling over into the world of jellyfish.

I found a seat at the U-shaped arrangement of tables and glanced around again. Of the seven people in the room, only one was a man. I'd never been to a scientific meeting like that before. Do jellyfish attract more women than men? I didn't think so. The majority of authors on the scientific papers I'd read were men. Most of the jellyfish experts I'd interviewed so far were men. Later I asked about the preponderance of women at the workshop. The scientists told me that it's not unusual now for graduate student classes to be weighted toward women in the natural sciences. Often the majority of students receiving PhDs are women. But, they added, many women leave their postdocs—the fellowships that academics need to complete before applying for university positions—for jobs outside of academia, so a disproportionate number of faculty are still men.

Internally, I raised my hand, acknowledging my personal contribution to maintaining gender inequity in science. I left academia during my postdoc too. After Keith and I were married, his salary won out over mine, and we moved to Austin. My postdoc position involved writing computer models of what satellites see when they look at the ocean. The staid lines of code had none of the grace of the mathematics I'd romanticized during my college years and felt completely divorced from the delight I'd felt when I fell in love with the ocean. As I trudged up the stairs to my windowless office on the second floor of a Cold War–era building to work in isolation, all I felt was dread. Thoughts of those years could still bring a knot to the pit of my stomach. Maybe that's why this elegant NCEAS building was so shocking to me. Computer models had come to equal cinder block in my psyche.

Even though I was miserable, the decision to quit academics came with a considerable amount of guilt. The scientific community had made an investment of nearly a decade in me. It had paid me, albeit frugally, during my training. And I walked away from it. A piece of me always felt that I left a debt unpaid.

A few more attendees joined the workshop on the second day, balancing the gender ratio a little. One was Monty Graham. My shoulders tensed at the sight of him. Our first conversation in Alabama had been so strange and had ended on such a sour note. Of all the interviews I'd done with scientists, his was easily the most confusing and awkward, so I wasn't sure how things would go now that I was seeing him again.

I found out pretty quickly. Monty entered the room explaining that he was late to the workshop because he was up against a proposal deadline. Then he noticed me sitting at the far end of the tables.

I waved hi.

He frowned. "We need to think about what we're going to do about her," he said, pointing a finger at me. "Are we able to speak freely in front of the evil journalist?"

Was he kidding? *I'm no evil journalist,* I thought. *Not even close.* Monty wasn't joking. I heard myself assuring everyone that I was not going to publish anything that wasn't already published. I wasn't there for a scoop, just to learn as much as I could about jellyfish ecology. Monty seemed sort of, but not completely, satisfied, and he excused himself to a side room to work on his deadline.

Rob Condon kicked off the workshop by talking about his work on the ever-so-glamorous-sounding topic of jellyfish slime. Slime is important throughout the biological world, especially in the sea. Slime is used for defense, to make swimming easier, and to capture food. When jellyfish form big blooms or when they're getting old, they give off a lot of slime, Rob said, but no one really knew how that affected the ecosystem. Working off the coast of Bermuda, Rob measured the chemical composition of jellyfish slime and found that it was mainly a lot of carbon-containing compounds, but not much else. Like a bottle of sugary soda, jellyfish slime is high-energy with practically no other nutritional value. Bacteria love the stuff. They slurp it up like sugar-crazed kids. With the extra energy, they grow like mad, multiplying and dividing at elevated rates. These fast-growing bacteria can outcompete the phytoplankton for nutrients, depressing the available food for many types of zooplankton that eat the ocean's unicellular plants. Fewer types of zooplankton mean less food for fish. The complexity of the food web collapses. Rob dubbed this newly discovered strand of the food web "the jelly shunt" because it shunts energy away from larger animals that we depend on for food and instead toward the growth of bacteria.

A researcher from Australia, Kylie Pitt, talked about jelly falls, which are just what they sound like: huge quantities of dying jellies cascading to the seafloor. I first heard of oceanic falls in the late 1990s from a scientist who recorded what happened when a gray whale died off the coast of California and sank to the seafloor a mile below. To the animals that live down there, a two-ton banquet had fallen from the heavens. These deep-sea creatures were used to scavenging on bits and pieces of organic detritus or, if they were lucky, a chunk of sinking fish or a stray crab claw. A whole dead whale was a once-in-a-lifetime feast that drew in revelers from miles around to share the bounty. Research into whale falls using submersibles has since revealed that some of those revelers, including hagfish and sleeper sharks, can eat over a hundred pounds of whale flesh per day. After the meat is gone, bone pickers called zombie worms suck the remaining marrow from the skeleton. Nothing goes to waste.

Another type of marine animal, called a salp, also can form huge aggregations in the ocean's surface waters. Salps look like a clear glass vase that opens on both ends. Some are just millimeters across and others about the size of your hand. They are about as closely related to us vertebrates—having complex nervous and digestive systems—as an animal can get without owning a spine. While a jellyfish has just one hole, the water passes all the way through a salp: in the top and out the bottom. As the water flows through the salp, it passes through a very fine mesh that sieves tiny plankton from it. When food clogs the mesh, the animal rolls the whole thing up and digests it. And then, spiderlike, the salp spins a new mesh. With this miraculous mesh, salps are able to eat the very smallest organisms in the sea, the tiniest phytoplankton and bacteria as well as much larger creatures like copepods. Salps can capture food that spans four orders of magnitude, which is like us eating a celery stalk as easily as a full-size tree. They are like the ocean's vacuum cleaners.

Salps are even more poorly studied than jellyfish because, besides being flimsy, they tend to be found offshore. They are often clumped together with jellyfish in ecological studies because they are similarly clear and bone-free. And like jellyfish, salp populations explode frequently, influencing the chemistry and ecology of massive swaths of the ocean.

Here's an example of what happens when a salp swarm gets cranking. In October 2008, ocean currents formed a circular eddy ten miles across, off eastern Australia. The spinning water acted like a straw, sucking rich, deep water to the surface, mixing fertilizer with light, creating a rich stock of phytoplankton. Unlike most jellies, salps are vegetarians, and so this eddy was a sumptuous buffet. Unlike most medusae, salps can clone themselves, and when food is abundant, their populations can increase by 250 percent a day. By the time the salps grazed down the phytoplankton, the eddy contained 40 trillion salps. If you had lined them all up, a report on the bloom noted, "they would have reached halfway to the sun."

When jellyfish or salps form huge blooms in the surface of the ocean, they eventually devour all the available food, or parasites find them, or they run into a water mass that's the wrong temperature, or they just get old and die. Then they fall to the seafloor in a big wad of goo. Sometimes the decomposing remains even form gelatinous lakes on the seafloor. And just as when a whale dies, it's party time for the creatures of the deep. This party is less of a full-blown banquet and more like cocktails and appetizers. Jellyfish and salps don't pack the nutritional punch of whales. Still, scientists have observed crabs, shrimp, sea stars, and fish grazing on fallen jellies. The denizens of the deep may not be the only beneficiaries of jelly falls either. Like all life, jellyfish and salps are carbon-based, so jelly falls are a carbon express train from the surface of the ocean down to the deep sea, where that carbon is stored out of reach of the atmosphere for millennia. It's thought that the oceans suck up about a quarter of the carbon dioxide we are adding to the atmosphere by burning fossil fuels. Adding to the ecological complexity, jelly falls and salp falls may be our allies in the fight to keep our climate in check.

After Kylie spoke, Monty Graham took a break from his proposal writing and rejoined the group. He wanted to present some work he'd done on the impact of jellyfish blooms on both industry and people. But he glared at me from the whiteboard. "I don't want you writing this down." He strode across the room and ripped my notebook from under my pen, folded it closed, and set it on the far side of the U-shaped table.

"Seriously?" I asked. "I can't take notes?"

"No," he growled. "This isn't published yet."

"All right," I replied, and leaned back in my chair to just listen.

Models are at the heart of science because they define the scientific method. Every hypothesis is really just a model of something in the world. And despite their simplifications and assumptions, computer models are one of the most effective tools we have for making predictions about the future. That's why many scientists use them so much.

Historically, jellyfish have been considered an ecological dead end, meaning they weren't important to the cycling of the ecosystem. Monty, for all his gruffness, is an outstanding scientist, and as outstanding scientists do, he questioned that paradigm. Earlier in the year, he had teamed up with some fishery biologists and sifted through all the available ecosystem models of fish for any that included jellyfish. They found under two dozen among hundreds that predict the future of fish in the oceans. Then they looked to see if those models predicted whether jellyfish benefited or damaged those ecosystems. The analysis showed a mixed bag. Jellies can benefit the microscopic plankton community by clearing out all the larger shrimplike feeders that dine on the tiny plankton. With their predators missing, those tiny plankton flourish. Jellies can also be bad for fish, both because they compete for fish's shrimplike prey and because they eat fish eggs and juvenile fish. But the scientists also found that jellyfish are good for some big fish and turtles because some of those larger animals eat jellies.

What eats jellies? It turns out, quite a few organisms. There are two well-known jellyfish predators. One is the sunfish, a behemoth related to the pufferfish; it eats jellyfish almost exclusively. It eats a lot of them too, because it often weighs in at as much as five thousand pounds. It has a shortened back end and is flattened side to side. If you Google videos of this amazing animal, you might be struck by its similarity to the floating head of a sumo wrestler. The barnacles that grow off the chin of this slowly swimming giant look like moles hanging off its thick face. The animal is often seen flapping its long fins at the surface, maybe to knock off some of these parasites or to encourage birds to pick them off as a snack. (And if you Google baby sunfish, you will be struck by its similarity to a Christmas ornament.)

Most sea turtles rely heavily on jellyfish, and the largest of the turtles, the leatherbacks, eat as much as 73 percent of their body mass in jellyfish every day, depending on the nutrient content of the jellies. In fact, turtle researchers use planes to spot blooms of jellyfish to find out where they should go to find turtles. But medusivory is not just for turtles and the odd sunfish. A study of the tissue composition of fish in the Mediterranean showed that jellyfish could account for about a third of the diet of several important commercial predators, including bluefin tuna, spearfish, and swordfish.

The interesting thing that Monty and his colleagues noticed in their ecological models was that there aren't a lot of animals that link the food web of the microscopic plankton to the food web of the larger ocean animals like fish. The fact that gelatinous creatures are big enough to be eaten by fish and yet have small enough stingers to eat the plankton puts them in an unusual ecological position—so unusual, in fact, that Monty and the other biologists wondered if jellyfish might even be a "keystone species."

The idea of keystone species was coined in the 1960s, and it refers to an animal that is so crucial to the function of an ecosystem that without it, the entire ecosystem collapses. The name comes from the center stone in an arch. Pull it out and the entire thing comes crashing down. One of the most famous examples of a keystone species is from the kelp forests on the Pacific coast. When sea otters were hunted nearly to extinction during the United States' westward expansion, the massive kelp were grazed to the nub, taking with them all the snails, fish, worms, and clams that make homes in the kelp forest. The culprits were sea urchins, which march across the seafloor on their tiny tube feet, munching on kelp roots. The sea otters, which eat the spiny echinoderms, had been keeping the urchin populations in check. The sea otters were the keystone that held up the entire kelp forest ecosystem.

Could it be that jellyfish—occupying a pivotal position between macroscopic and microscopic food webs and generating unique bloom-and-bust cycles—were critical for a well-balanced ecosystem? Instead of seeing jellyfish as a signal of ecosystem demise, Monty and his colleagues wondered if jellyfish might actually be essential, if they might actually be "the most important predators in the sea." They found some evidence that they might, at least in some places.

In the Chesapeake Bay, jellyfish called sea nettles exist side by side with comb jellies. In the spring, the bay is full of oyster larvae, which provide rich nutrition for fish. But the comb jellies are so efficient at feeding on oyster larvae that there's nothing left for the fish. That is, unless the sea nettles are around. Sea nettles could feed on oyster larvae too, but they don't. If you watch a sea nettle ingest an oyster larva, a few seconds later you'll see it spit the oyster larva out. (Anyone who doesn't like raw oysters can sympathize.) What the sea nettles do have a taste for is comb jellies. So the sea nettles keep the system in balance by culling the comb jellies, leaving more food for fish, and allowing the oyster population to reseed itself each year. When the sea nettles disappear, the fish struggle to compete and the oyster population crashes.

In other places around the world, like the Bering Sea, the northwest Mediterranean, and the Norwegian Sea, Monty and the other biologists discovered that jellyfish might be abundant, but given the data we had, they didn't seem to play such a crucial role in the ecosystem.

The study led to some new hypotheses. Jellyfish seem to be more important in the ecosystems of enclosed or partially enclosed bays than in open oceans. Also, it seems that the more an ecosystem is exploited by human activities, the larger the role jellyfish play in it. Often these two factors go hand in hand. Humans have long settled enclosed bays because they afford security for ships and coastal structures and yet give access to the sea's resources. Along with sharing new research on jellyfish ecology, one of the major goals of this workshop was to test these hypotheses about jellyfish and their importance in different ecosystems.

The scientists at the workshop agreed that the fisheries models that Monty had studied greatly oversimplified the role of the jellyfish. Lumping the many different species and sizes of jellyfish together into one ecological category just doesn't do the jellies justice. In any ecosystem, there might be dozens of species of jellyfish, each with particular preferences for food and water temperature, different abilities to swim and avoid predators, not to mention various life cycles and growing seasons. So the group wanted to build a model with more complexity, starting by recognizing the differences between gelatinous animals that live the polyp-medusa lifestyle and those that don't.

Tammi Richardson, a skilled ecosystem modeler, headed to the whiteboard and started writing down the outlines of the ocean's food web, the boxes and arrows that show the way energy and carbon flow through the biology of the sea. At the same time, the other members of the workshop dug into the scientific literature—and they recruited me to dig in too—to find published values of population numbers, feeding rates, growth rates, and respiration rates: the numbers that would fuel the computer-generated food web. By the last day of the workshop, we had collected enough of the required values for Tammi to run the model she had been programming.

Tammi said, "Before we do this, I need to do the 'dance of luck.'" She stood in the center of the U-shaped array of tables, kicked her heels in the air, and did a jig. A pretty good one too. Then she spun back around to her laptop and pressed ENTER. The results flashed up on the overhead projector.

"The dance worked!" Tammi said. The model had run successfully, meaning that the inputs were solid, if preliminary. This effort was the start of a continuing process to add the complexities of jellyfish ecology to our understanding of the sea.

Since then, jellyfish scientists have continued to chip away at the ecological block into which jellyfish have been traditionally lumped. A few years after the meeting in Santa Barbara, Kelly Robinson, Monty Graham, and several colleagues published work analyzing ecosystem-based models of fisheries that included jellyfish, work that began at that meeting. They quantified the ecological importance of animals by their *reach* and *footprint.* A population's reach is the amount of energy that it transfers up to the predators that eat it. Its footprint is the amount of prey energy required to support the population. The work focused on three locations: the Bering Sea, the Gulf of Mexico, and Northern California. In all three cases, when the scientists created the models with more jellyfish than small prey fish, the bigger predators declined. That means fewer tuna, dolphin, and seabirds. However, the researchers wrote, to think of jellyfish as an ecological dead end was the wrong traffic metaphor. Instead, jellyfish are more of a roundabout, shifting energy from going straight up to the top predators and distributing it to animals throughout the food chain. In comparison, small fish like sardines and anchovies are an energy expressway feeding the large predators. The study pointed a finger at a crucial part of the food web, the small fish.

Today, a full third of the fish we pull from the sea, 31.5 million tons, are small schooling fish. And we don't even eat them. These massive quantities of prey fish are used in pet food, as feed for farmed chickens and pigs, and to make fish oil vitamin supplements. By weight, six times more small fish are fed to farm animals—animals that because they live on land wouldn't normally eat marine fish—than the seafood eaten by the entire population of the United States. The models showed that as we remove more and more small fish from ecosystems, the jellyfish create a more circuitous food web. The balance between these small fish and the jellyfish is where we should be looking if we want to understand our future seas.

Although there were still a few hours left before the workshop adjourned, I had to leave to catch my flight back to Austin. I packed up my computer and papers, as well as my reporting notebook that I had retrieved from where Monty had banished it at the far end of the table. I made my way around the room, thanking everyone for allowing me to join the meeting, though Monty Graham continued to avoid me.

As I said good-bye to Tammi, she pulled me aside and said, "I'm chief scientist on a cruise off of Bermuda in July that's focusing on jellyfish. Rob's going to be there too. We are going to go blue-water diving beforehand to survey the jellyfish. Let me know if you'd like to join us."

"Are you kidding?" I asked. It was my turn to do a little jig, in my heart anyway. "I would love to!" As I left the conference room, I looked back at the U-shaped tables and the large white screen. Tammi's model run was still projected there. The web of boxes and lines suddenly bore an uncanny resemblance to a grinning face. I was going back out to sea!

Back in Austin, I buzzed around, planning for the Bermuda trip. I wrote a grant proposal to pay for my flight, called the local dive shop to schedule a refresher course in scuba and to find out the fastest way to rack up enough dive hours to qualify for certification at the Bermuda lab. I set up after-school activities for my kids, babysitting, carpools.

But then the cruise dates were shifted. No problem. I shifted my plans too. Rescheduled the dive classes, babysitting, carpools, flights.

And then, really bad news. Because of schedule conflicts, Rob canceled the blue-water diving portion of the cruise entirely. Tammi told me she was still going out to sea on the ship to study the plankton, but not jellyfish specifically. She assured me that I was still welcome to come along. I made the tough call not to waste the money and time or put the stress on Keith and our kids if the research wasn't going to focus on jellyfish.

As the dates of the cruise came and then passed, I fell into a funk, complaining to everyone I knew. One of those people who received the brunt of my sulking was a friend who is a novelist. She mentioned that she was thinking of writing a screenplay and was studying the narrative structure of movies. A lot of movies' plots are strikingly similar, she'd learned. The protagonist, she said, always takes a journey sixty-one minutes into the movie. "You can time it. *Star Wars*, *Toy Story*, *How the Grinch Stole Christmas . . .*"

That didn't make me feel any better. I moped. Like a slippery medusa, my jellyfish journey was sliding out of my grasp.

At two in the morning, I woke with a start. My head was spinning. I sat up in bed and threw off the covers. "I haven't lost the journey. I'm already on it!" I made this pronouncement to a slumbering Keith. He rolled over, unimpressed.

But I was inspired. I grabbed the iPad from where I'd stowed it in my nightstand and popped open the cover. The glare filled my pupils as they contracted. I frantically pecked at the virtual keys, writing down each of the moments in my life that had brought me to this point. I was a mid-forties suburban Austin mom with an obsession with jellyfish and their role in the world. How the hell did that happen?

It wasn't random. There was an answer to the question.

I wrote a list of scenes, important moments in my life, the events that had led me here. There was a pita cart scene from Israel, before I pulled the advertisement for a marine biology course off the wall. One from college, when I resolved to study math despite being told that women shouldn't be studying math. One from Woods Hole, when I worked for a cruel boss in a marine ecology lab. One after college, when I failed miserably at being an accountant. I listed moments in grad school in L.A., when cockroaches roamed my apartment and rats fell from the ceiling tiles. I wrote about being with the wrong guy and not recognizing it. About meeting Keith, and not recognizing it. I wrote about nearly losing him. I recalled the terror of a flight gone wrong; about soul-sucking days as a postdoc.

And when I looked back over the moments, I noticed a trend. Not all, but most, were really low points in my life. They weren't dreadful moments. They were fairly ordinary. I recognize that I've been privileged and blessed by my birthplace and family, by education and opportunity. My suffering is minimal compared with that of so many others. But in the middle of this restless night, I was staring at a complete list of my bad times. The troughs of my waves. They were times when I was lost and boxed in by circumstances out of my control. When I felt confused and unable to make sense of the world. When it seemed I had no choices. And then I saw, looking back, that these low points were actually turning points. They were the moments that pushed me to make a change in my life. To break up with a destructive boyfriend. To change my career. To change it again. To move from a toxic city. To choose the right person to marry. The low points—not the moments of joy—were the decisive moments in my life. I had almost never made a change when things were going well. Why would I, if I was happy and comfortable? It was only when things got so stressful, so difficult, or so miserable that I couldn't stand it that I took action. I had to *feel* bad to fix my life.

Because I was so infused with jellyfish and the ocean's problems, I saw the same logic applied to the world around us. And I suddenly got it. For most of us—at least those of us lucky enough to live in developed countries—the environment doesn't seem that troubled day-to-day. We are mostly comfortable in our world. Our everyday experience with the environment is functional. Our world supplies us what we need and expect. In the summer, we can bet on blue skies and green grass; in the winter, icicles glitter and geese fly overhead. The ecosystem that brings us grass and geese supports us. Our stores are filled with fresh food. Our seafood counters are loaded with seemingly endless assortments of fish and shrimp, scallops and lobster. When we breathe in, we don't choke. Our water, with few exceptions, doesn't make us sick.

That's not always the case, of course. Intermittently, we have been receiving brash and more frequent reminders that things are not as they should be: heat waves, droughts, fires, and floods. But devastating storms and floods that upend our stability are still thankfully rare. Most of the time, those of us on the industrialized side of the planet are comfortable. Our world still works.

So why would we change the way we treat the world? Why isn't protecting our life-sustaining ocean more urgent? We haven't yet reached a low point. Intellectually, we know there are a lot of problems, but we don't yet feel them intensely enough.

I was reminded of a moment in the phone call I'd had with Jenny Purcell when I first started with jellies, one that'd made a lasting impression on me. She'd told me that if I wanted to write a book that mattered, I needed to tackle the bigger problems on our planet, like climate change, ocean warming, and ocean acidification, all of which stem from our unchecked use of fossil fuels.

"Just to play devil's advocate," I'd retorted, "you could say, Look how much we've accomplished by developing fossil fuel markets. We've improved the global standard of living. We've made technological advancements, scientific advancements—"

"Well," Jenny interrupted, "we are an amazing species. Capable of many amazing things . . ."

Her voice trailed off, but I knew what she had left unsaid then, and it came flooding back now. This wasn't about being capable; it was about being willing. Making a difference would require our collective intention and effort. It would mean growing a spine.

Ephyra

Stop Waiting

Most jellyfish live a life in two parts. The first part, the polyp, lives attached to a solid surface. It stretches its tentacles out to ensnare its prey and then hauls them in toward its mouth. The polyp can scoot a little, moving beyond its home to a slightly more favorable habitat. After a while, for reasons that are still mysterious, the jellyfish polyp prepares for the second part of its life. It changes its form, schisming into a stack of nascent baby medusae, each one capable of surviving on its own. When the time is right, these ephyrae swim free into the great, open ocean to search for sustenance on which to grow. From my base in Texas, I had explored the biology of jellyfish; the animals' history both ancient and modern; their genes, movement, glow, senses, nerves, and ecology. Finding these stories was like stretching tentacles and encountering nourishing treasure, all the while moored in the safety of home. But after my feverish revelation about my life and the ocean, I schismed too. The time had come to leave the safety of the surface that supplied support and comfort.

As I was living in Texas, I had the urge to go big. Which could mean only one thing: the giant jellyfish that swarmed the seas of Japan. Giant jellyfish go by several different names. In English, they are Nomura's jellyfish; in Latin, *Nemopilema nomurai*; and in Japanese *echizen kurage. Kurage* is the Japanese word for "jellyfish"; *echizen* refers to the province where the animal was first reported. Although the lion's-mane jellyfish is usually considered the biggest jellyfish by length, the giant jellies are tops for weight, tipping the scales at nearly five hundred pounds. Think baby grand piano.

In the twentieth century, large numbers of giant jellyfish were observed off the coast of Japan about once a generation: in 1920, 1958, and 1995. But then they bloomed just seven years later, in 2002. And again in 2003 and in 2004. In the past decade, giant jellyfish have appeared every year, although their numbers do fluctuate significantly.

In 2005 there was a record-setting bloom, with as many as half a billion blimplike invaders passing through the strait between Japan and South Korea each day. Besides clogging fishermen's nets, the giant jellyfish swarm reduced the fish catch and poisoned the fish. The Japanese Fisheries Agency fielded complaints about jellyfish from more than 100,000 fisherman, and damages were estimated in the range of 30 billion Japanese yen, or $260 million. But 2009 broke the record again, with an increase in giant jellyfish abundance of about 25 percent over 2005. The weight of these megamedusae caught in fishing nets even sank an ill-fated boat, tossing the crew into the sea, though luckily, no one was injured.

To find out more about the giant jellyfish, I wrote an e-mail to a scientist I had spoken to about jellyfish polyps, Lucas Brotz. He had studied populations of jellyfish around the world, and I thought he might have connections in Japan. I asked if he knew of any Japanese scientists whom I could ask about seeing the giant jellyfish. The very next day, Lucas wrote back that he was planning to go see them himself. He was attending an international conference on jellyfish in Hiroshima and then going on a dive trip. One of the foremost Japanese jellyfish scientists, Shin-ichi Uye, had connected him with a dive shop located near a place where giant jellies were often seen. Lucas suggested that I go to the meeting and take the dive trip too.

I knew Shin-ichi Uye's name from journal articles I'd been reading. He had been working to tease apart the conditions, like wind speed and water temperature, that correlate with giant jelly blooms. With the Japanese Fisheries Agency, he had been developing an early-detection system to warn fishermen about an approaching front of giant jellyfish in nearly the same way that meteorologists predict rainstorms. To ground-truth his models, he and his colleagues rode ferryboats, scanning the water surface and counting jellyfish. My Google alerts had brought me a recent news article from Japan reporting that researchers on the ferry surveys had spotted two thousand times more jellyfish than at the same time last year. I figured that there would be a big bloom of giant jellyfish. I really wanted to see it, and now I might have a way in.

I was too excited to wait for Keith to get home from work, so I asked him to meet me for lunch. He suggested a Chinese restaurant halfway between his office and our house. Before we even sat down and ordered, I told him about Lucas's e-mail, the international conference completely dedicated to jellyfish, and the potential dive trip. It had been a decade since I had dived in the sea, but I had already started a refresher course for the canceled Bermuda trip.

"Do you think you can take care of the kids?" I asked. "I've never left for this long. I'll need to be gone almost two weeks. Do you think the kids will be okay without me?"

"We'll probably all fall apart by the time you return," he said, taking the paper off his wooden chopsticks and breaking them in half. "The kids will have failed out of elementary school, and I'll have run off to Vegas with the dog."

After we finished eating, Keith handed me my fortune cookie. I cracked it open and read the words on the little piece of paper, then silently handed it to him. It read: "Stop waiting! Buy that ticket. And take a trip today."

"I guess you're going to Japan," Keith said.

That night, after the kids were in bed, we used up fifty thousand of his frequent-flier miles. I would fly from Dallas to Tokyo, then take the bullet train to Hiroshima.

As I brushed my teeth, I made lists of things I needed for the trip: lithium batteries for my underwater camera, decent shoes for the meeting, audio recording software for the iPad. As I watched eggs harden in the pan and then slid them onto plates with just the right amount of salt so Ben wouldn't complain, my thoughts swam to the dive class I had signed up for. As I put lunches in backpacks and zipped them up, I thought of flight confirmations and train schedules. As I kissed the tops of heads running out the front door, I thought I'd better print out hotel and dive-shop addresses so I'd have hard copies. And then I sat down at my computer and opened an e-mail from Lucas:

> **Hi Juli,**
>
> **See text below from Shin. Not looking good. I'm not sure what to do at this point . . .**
>
> **Lucas**
>
> **Hi Lucas:**
>
> **In regard to your plan of SCUBA with *Nemopilema*, we have been communicating with SeaMore dive shop staff about current appearance of medusae. At this moment, jellyfish is very scarce, only a few medusae were caught in trawl and set nets in their area. No SeaMore staff has encountered *Nemopilema* this season. Therefore, I recommend not to make to the Sea of Japan.**
>
> **As I mentioned in my early emails, *Rhopilema* is in a big bloom in Ariake Sea, so that going to Kyushu is more promising.**
>
> **Cheers,**
> **Shin**

I was crushed. This is what jellyfish scientists told me they struggle with all the time. When you plan to go study them, they are nowhere to be found. When you can't get away from your office, they bloom like crazy. Jellyfish are just flat-out unpredictable.

The second jellyfish Shin mentioned, *Rhopilema esculentum*, is one of the most important commercial jellyfish. Smaller than the five-hundred-pound giant jellyfish, these jellies grow to only around seventy pounds, which is still a massive amount of gelatin. And they're beautiful, with turquoise bells and brick-red undersides. The species has been prized as food for hundreds of years in China. It's also used medicinally to treat high blood pressure and bronchitis. It's so desired a commodity that populations have been depleted—one of the few examples of jellyfish overfishing (another perhaps being the *Aequorea* jellyfish in Puget Sound, used for green fluorescent protein). To combat overfishing, scientists in China have worked with aquarists and fishermen to seed the ocean with baby jellyfish, with the aim of stabilizing a naturally variable fishery. In one year alone, they released 400 million dime-size juvenile jellyfish into the northern Yellow Sea. The seeding boosted the jellyfish yield by over a third. The enterprise makes good financial sense; the ratio of the cost of seeding to the value of sales was an impressive one to eighteen. In 2006, the profit to an individual fisherman was six hundred yuan, or around eighty dollars. When I spoke to Jenny Purcell at the beginning of my jellyfish research, she was in China. She confirmed that the jellyfish seeding operations may be growing: "It's very economically valuable. They can put a little money into aquaculture and get back a lot. Other countries want to do it as well, and it may have already spread to Southeast Asia."

Now a shallow bay in southern Japan, the Ariake Sea seemed to be teeming with these jellyfish, and in another e-mail from Shin, I learned that something of a jellyfishing frenzy was under way. Boats had been switching over from shrimping and traditional fishing to jellyfishing. A three-person crew—a captain, a scooper using a fifteen-foot bamboo pole with a net at the end, and a cleaner—could catch three hundred medusae a day. For their efforts, the crew earned about $1,800. The local fish market was able to process about ten tons per day, but the fishermen were bringing in much more than that. An additional sixty tons per day were being exported to China. Shin expected that the jellyfishing would continue through the timing of my visit. The commercially important jellyfish seemed like a good backup plan, though not nearly as exciting as the giant jellyfish.

I woke up with two spots of pressure pushing down on my biceps and a warmth against my chest. I became aware of the sound of a soft, deep feline motor. Our cat, with a reputation for being grumpy and aloof, affectionately kneaded my upper arm with her paws. My awareness groggily crept away from the central core of my body to the other end of my arm. My fingers were entwined with Keith's, our connected hands thrown across his bare chest. How was I going to leave this place? How was I going to pick myself up from this comfortable bed and leave my children behind? I'd never left them for this long, and now I was going to get on a plane to hunt for jellyfish in Japan. Was I crazy? I wasn't some sort of *New York Times* correspondent or a Discovery Channel celebrity. I didn't have a job as a staff reporter for *National Geographic*. I didn't have a producer or a crew or a budget. I didn't have a fixer or a scheduler. I was using frequent-flier miles and money out of our pockets to go chase jellyfish. I didn't know how to read one kanji. I couldn't say anything in Japanese except the names of some fish from sushi menus and *Dōmo arigatō* from the Styx song "Mr. Roboto." I didn't even know if I'd actually see jellyfish at all, much less a giant one, or even a commercially valuable one. What was I thinking?

I walked out of the train terminal into the muggy night air of Hiroshima and headed toward a lineup of taxis, then pantomimed to the driver of the first one that I'd like a ride. He rushed over, grabbed my backpack, placed it in the impeccably clean trunk, and gallantly opened the door. I slipped inside a space remodeled to look more like a living room than a Toyota. Hand-sewn and -piped lace covers sheathed the seats. A small shrine was glued to the dash next to a television tuned to a baseball game. Tissue boxes with matching lace hung from the back of each headrest. When we arrived at my hotel, the door magically popped open for me. I had arrived, if not full of confidence, at least in style.

Once in my room, I dropped my stuff on the floor, brushed my teeth, drank a bottle of water, and then collapsed onto the bed. I picked up the pillow to give it a fluff, but rather than kneading my hands into cushion, I discovered a hard beanbag of rice. As I lay my head down on the unexpectedly comfortable surface, it barely made an indent. I started to compare my current resting spot with the last bed I'd slept in, my hands entwined with Keith's and the cat kneading my arm, but before I could complete the thought, I was asleep.

Even though I'd worked just blocks from Hollywood for a decade, I never cared much about the celebrities. But at this meeting I was starstruck. It was full of my kind of luminaries, giants of the jellyfish world. Monty Graham was the first to speak. He was followed a few talks later by Jenny Purcell. Next was Shin-ichi Uye, who I hoped would help Lucas and me cobble together some sort of jellyfish expedition. He was a distinguished man with an uncommon combination of sharp intellect, modesty, and boundless energy. I would come to think of him as a philosopher-scientist. From my seat near the back of the conference room, I noticed a guy who had to be Lucas, based on the photo I'd seen on his University of British Columbia webpage. In that picture, Lucas was bundled in a parka and a heavy, white snow hat. Without the winter bulk, the real Lucas was thinner and balder. When he got up to give his talk on human impacts and jellyfish blooms, I saw that he was taller too, over six feet. Lucas spoke with poise and confidence and kindness, and I immediately felt that we'd be okay traveling together.

A Japanese scientist named Reiji Masuda took the stage. He turned to the audience and got a laugh by declaring, "I am interested in laboratory experiments on fish behavior so I call myself a fish psychologist." Masuda showed a video of his laboratory, a warehouse filled with large aquaria, each about the size of an aboveground swimming pool. Fish swam in lazy circles in these tanks, and Masuda said he performed experiments comparing how they fared when fed a regular diet of krill with how they did on a diet supplemented with jellyfish or a diet of jellyfish alone. Like a human psychologist who understands that each individual comes to treatment with a different background and different predilections and aversions, Masuda said that each species of fish was equipped with different adaptations to the environment and responded differently to his experiments. Filefish could survive on a diet of jellyfish alone. Jack mackerel wouldn't eat jellyfish unless they were starving, but they did swim with them, hiding in their tentacles for protection. Red sea bream and tiger pufferfish couldn't survive on jellyfish, but the jellyfish did provide important nutrients. Fish that ate a diet that included jellyfish contained healthier fatty acids, including more omega-3 fatty acids, than fish that ate no jellyfish. Research like the kind that Antonella Leone is conducting in Italy indicates that jellyfish are high in these fatty acids, which are an essential component of our cerebral cortex, skin, and retina, not to mention sperm and testes. They are being studied for potential positive impacts on people with Alzheimer's, cancer, ADHD, and during pregnancy. "So," concluded Masuda, "feeding jellyfish to fish is good for the health of the fish, and good for our health too." I could picture the next big thing at the meat counter. If not jellyfish themselves, near to the grass-fed beef we might find jellyfish-fed mackerel.

A bit later, a researcher from Hiroshima University told a tale of a small pier installed in Hiroshima Bay. Its dimensions were about the footprint of a three-bedroom house. The underside of the pier was monitored with cameras for moon jelly polyps, which appeared after four months. A month later, the entire underside of the pier was carpeted in polyps. These polyps released 23 million baby jellyfish ephyrae into the Bay. A cubic meter of seawater held between one and six ephyrae. Afterward, those numbers tripled. All from one dock the size of a modest house.

At the end of the talk, I raised my hand to ask whether there are ways to mitigate polyp growth when you install a new pier. Unfortunately, my question was lost in translation, but later, Reiji Masuda came over to talk with me. He had two ideas. "And they are both very simple," he said. One was to drill holes in the pier to let light through. In Tokyo Bay, moon jellyfish polyps seem to produce more medusae in the dark. They live on the undersides of things, in the shadows. It's possible that if there's enough light, they won't produce so many medusae. His second idea was to encourage predators to eat the polyps.

"What eats polyps?" I asked, having actually never thought about it.

"Nudibranchs love them." Nudibranchs are sea slugs, shell-less snails that roam rocks and seaweed stalks, ingesting the soft, sessile creatures they crawl over. Unlike the ugly brown creatures that the name sea slug might have you imagining, nudibranchs are some of the most vibrantly colorful creatures you'll find. Many have tufts of frill across their backs that make them look like living feathers. "But nudibranchs cannot swim, so we must provide them with a way to get to the polyps," the fish psychologist said.

He really does try to get into the mind of animals, I thought. The fish psychologist explained that if a pier were built with cables that reached to the seafloor, nudibranchs could climb them the same way they climb a kelp stalk. When they reached the underside of the pier, the nudibranchs would munch away at the jellyfish polyps. Much as we use ladybugs to keep our tomato plants free of aphids, sea slugs could act as a natural polypocide.

Sea slugs have fascinating digestive capabilities. When they eat, not everything gets turned into a slurry and then broken down to its constituent molecules as it does in our guts. Instead, sea slugs somehow preserve entire functional units of cells from the corroding effects of their digestive juices. They then reinstall those units in their own bodies, essentially hijacking cellular equipment from their prey. Several nudibranchs have the ability to salvage the chlorophyll-laden parts of the plant cells they eat and then install them in the lining of their gut, giving them a green tattoo. The practice has given the creatures the nickname "solar-powered slugs," but it's not yet clear whether the creatures are able to actually harvest the energy from the sun produced by their photosynthetic theft.

Other brightly plumed sea slugs called aeolids target a different kind of loot: the stinging cell. These sea slugs eat jellyfish and their kin and then plunder their cellular weaponry. The stinging cells pass through the animal's gut and lodge at the ends of feathery extensions that line their backs. When an unwary predator approaches, the stinging cells fire just as they do in a jellyfish. One of these stinging-cell-bearing slugs (*Glaucus atlanticus*) is called the blue dragon because the six azure extensions that splay like wings off its body look like the wings of a mythological reptile. It lives upside down, floating on the world's largest surface, that between the ocean and the air. Its blue side faces up, disguising it from birds. Its silver side faces down, making it invisible to fish. Among the prey whose stinging cells it steals is the Portuguese man o' war, which contains some of the fiercest and most poisonous stinging cells in the sea.

Dena Restaino, a graduate student from New Jersey, recently had the idea that this stinging-cell looting might just help solve one of the most pressing questions in jellyfish research: Where are all the polyps? Dena collected sea slugs from Barnegat Bay, which has a somewhat serious problem with stinging sea nettles. She analyzed the sea slugs, not for sea slug DNA, but for the jellyfish DNA in the hijacked stinging cells. And she found it. The slugs' stolen DNA reported that they'd been eating mostly the polyps of two species of jellyfish not known to cause stings. But about 11 percent of the jellyfish DNA was from the stinging nettle, confirming that the entire life cycle of the jellyfish exists within the bay. In the future, sea slugs might be able to act as informants, telling us not just where the polyps are, but which species are hiding on our docks and jetties and how those populations are changing.

After the meeting ended for the day, I saw Lucas motioning to me from across the room. He was speaking with a Japanese scientist, and they were peering intently at an iPad, looking at a map of Japan. Red dots showed where giant jellyfish had been caught in fishermen's nets in the last couple of days. China was on the west side of the map; the Korean peninsula hung down the center; Japan was to the east. The scientist explained that giant jellyfish are born in the Yellow Sea between China and Korea. They are swept south around the Korean peninsula in the Tsushima Current. Clearing the southern tip of Korea, the current turns north and splits into three branches. One sweeps east along the coast of Japan, one goes out into the center of the Sea of Japan, and one curves back toward Korea. This year, the giant jellyfish were in the arm that curves toward Korea. Disappointingly, none of the jellyfish were near the coast. But there were some red dots near tiny islands in the center of the Sea of Japan.

"What are those?" I asked.

"Those are jellyfish caught by fishermen who live on the island of Tsushima."

"Why don't we go there?" We had four days to kill before Shin could help us get to the Ariake Sea for the commercial jellyfish, more if that trip didn't work out. We definitely wouldn't be seeing giant jellyfish sitting around Hiroshima.

"No, I don't think you can go there," replied the Japanese scientist. "You cannot find anyone to speak English there. You cannot see the jellyfish from the shore, so you must go out in a boat. I think it would be very hard."

Damn, I thought.

But Lucas said, "Let me talk to Shin. Maybe he has some ideas."

After the meeting ended for the day, jellyfish scientists lingered in clumps, making dinner plans. I followed along with a group that exited the hall and wandered together on the manicured streets of Hiroshima. The Japanese scientist with the iPad was part of the crowd, and he offered to act as guide and translator. The entrance to the restaurant he chose was up a few worn steps, and we all ducked below a half curtain into a cramped foyer as we entered. After removing our shoes and stowing them away in wooden shelves as though we were going to a yoga class, we sat on pillows on the floor, our feet dangling into a hole below the table. When Monty Graham took a seat to my right, I cringed a little. What zingers was he going to toss at me this time? We ordered drinks, cocktails made of Japanese plum liquor and potato vodka that blended beautifully with the sepia wood and the dim light.

The Japanese scientist ordered food for the whole table, mixtures of fish and vegetables, sushi, and pickled greens. We non-Japanese speculated on their origin and how the dishes were prepared. Stories of research cruises and rough seas were exchanged around the table. Encounters with frightening creatures, both spineless and skeletoned, were shared. Monty told us a story involving a ditch, some guns, a pickup truck, and the rescue of an enormous alligator turtle that was about to become someone's dinner near his lab in Alabama.

After the conversations broke into small groups, I turned to Monty and brought up the Discovery Channel documentary that featured his dissection of a giant jellyfish. "What was it like to dive with giant jellyfish?" I asked him.

"Honestly, we had a hard time finding them. When we got over here, they were nowhere to be found." Monty said that the film crew finally located one animal in a fisherman's net in South Korea, and they told him to keep it alive. "We flew there and did all the shots of that one jellyfish."

"Sounds familiar," I said. I explained that Lucas and I were originally hoping to dive with giant jellies, but now we were just hoping to see them. Even that was looking unlikely.

With our extra time, Monty suggested that we take a day trip to Itsukushima, a nearby island popularly called Miyajima (Shrine Island). It was considered a sacred place and had lovely trails and old Buddhist temples.

I was reminded that in Santa Barbara, Monty had also found a great side trip—that time to an artist's studio—although I didn't join him. Then, my tongue loosened by the liquor, I said, "Back in Santa Barbara, you were not much of a fan of me. You even called me an evil journalist."

"Back in Santa Barbara, I didn't trust you. I did think you had an agenda. But you've followed up. You've shown that you aren't here just for a quick story."

I leaned back against the wall of the booth savoring that comment. While I don't think I had ever taken the slur "evil journalist" seriously, it still stung. And the acknowledgment felt good. Don't we all hunger for approval? More than that, Monty's comment felt like a validation of my efforts to follow my passion for jellyfish. I was here in this restaurant on a twisty street in Hiroshima at a table full of jellyfish scientists not because I had drifted here, but because I'd made it happen.

Sacred Island

On the final day in Hiroshima, a group of the jellyfish scientists invited me to join them on a day trip to the sacred island Monty had mentioned in the restaurant. We disembarked from the short ferry ride near a temple that appeared to float on the sea. It was guarded by a huge reddish Shinto gate. If you've seen only one photo of a Shinto gate, this is probably it. The structure is located several hundred yards offshore, but when the tide receded later in the day, it cleared a soggy sand path, allowing visitors to hike up to its massive legs and squeeze coins into the cracks in exchange for a wish.

Ancient trails wandered up the mountain in several directions. We chose one that followed a gurgling stream. Nooks and crannies in the rocks contained small stone shrines tucked into extravagant foliage. Pleasant music was even piped in from hidden speakers. Our climb brought us to an imposing wooden temple from the early Edo period, about four hundred years ago. We removed our shoes and slipped on plastic sandals from a bin on the temple's steps. Inside the building, which was only partially enclosed from the elements, rough-hewn rafters doubled as shelves for ancient wood paintings of samurai and dragons. Next to the temple a red five-story pagoda soared upward, dwarfing the evergreens.

Beyond the temple, the hike grew more rugged. Every so often we had to pull over to let a line of schoolchildren dressed in matching primary-color shirts and baseball caps pass us on their way down. They held their hands in the air to receive high fives from us. At the top of the mountain, we climbed a rusty set of stairs to reach the platform of a lookout tower. The Inland Sea glimmered in the sunlight. Across the bay, the buildings of Hiroshima encrusted the coastline like barnacles laying claim to the flattest parts of the valleys. Mountainous islands in graduated shades of blue filled the horizon.

The sea itself was very quiet. There were no sailboats or ski boats. No jet skis or even fishing boats. Only a few ferries followed the well-worn course back and forth between Hiroshima and the island. The route cut through rows of parallel floating piers in arrays of three or four by five or six. These raft clusters stretched as far as we could see. They were used for farming the sea, many for growing oysters. Here was a place where the oceanic Garden of Eden that Ferdinando Boero had spoken of, a place where we were still hunters and gatherers, had been tamed.

As we turned and walked back down the mountain path on Miyajima, Lucas and I talked about the view from the top of the hike. He said, "You know, we're on the cusp of no longer having a wild ocean. We may not have any more wild fish soon. But we'll still need to eat fish from the ocean." He was referring to the fact that the oceans are the world's largest protein source. According to the UN, more than 3.1 billion people depend on fish for nearly 20 percent of their protein. The absence of fish protein would leave a crippling vacuum for many people in the world. "The ocean will be like on land, where the food we harvest isn't wild anymore," Lucas said. "We've agriculturalized our food sources on land, and we'll have to farm it all from the sea as well. Our future may be that Inland Sea, all covered in farms."

I thought about a future in which the sea is no longer wild. Is that kind of future wrong? I think so. The great herbivores, like bison and elk, no longer roam across the land. Instead, their cousins, the cows, live in stockyards where they are part of a great agricultural machine. In my neighborhood in Austin, deer deprived of habitat and natural predators graze front lawns down to the nub. Human infringement on nature and domestication of plants and animals has to occur on land, because we live there.

We don't live in the sea, so we have the physical space and the opportunity to protect our oceans' future. One tool we have is marine protected areas, or MPAs, places where human activity is restricted, as it is in national parks on land. The idea isn't new; it's just been practiced inconsistently. In 1903, Teddy Roosevelt, with the support of the Audubon Society, established the Pelican Island National Wildlife Refuge in Florida to protect birds, which were being killed for their feathers so that ladies could decorate their hats with plumage. Today, there are over 1,700 marine protected areas recognized by the United States, but they form a complicated mosaic. Marine protected areas can range in size from tiny, just a couple of square miles, to massive. Just four MPAs near Hawaii and around remote islands in the Pacific cover as much space as California, Oregon, and Washington combined.

Only three percent of the ocean is protected, a fraction of the ten percent that experts project we need. But there's momentum toward more. In 2016 mega-MPAs were established off Chile, Palau, the Pitcairn Islands, and Saint Helena in the South Atlantic. Also that year, before he left office, President Barack Obama quadrupled the size of the Papahānaumokuākea Marine National Monument off Hawaii and established the first marine protected area in the North Atlantic. A few months later, fishing gear that damages the seafloor was banned from a region the size of Virginia that stretches from Rhode Island to North Carolina to protect deep-sea coral that can take as long as California's redwood trees to grow. And Mexico set aside 160 million acres of sea, or 23 percent of its marine area.

Even though marine protected areas are often compared to national parks on land, the protections they provide are highly variable, and for the most part nowhere near as comprehensive as what you think of when you think national park. In the vast majority of marine protected areas, the protection means some limit to human activities. However, that limit can be minimal. It could mean that fishermen are not allowed to drag nets across the seafloor or that boaters can't throw out an anchor where there's coral or seagrass. Fishing, even commercial fishing, is often allowed. So even though the percentage of protected area is headed in the right direction, regulation and enforcement remain somewhat hazy.

One potential reason for the growth in MPAs is that, although it seems counterintuitive, cordoning off parts of the sea from fishing actually increases fish yields and profits for some fisheries. When it is protected from fishing, an MPA maintains a healthy population of reproductive-age adults. As most fish release many hundreds or even thousands of eggs when they spawn, they produce more offspring than can live in the protected area. The fish stock spills over into the fishing grounds.

Unlike on land, where the bulk of the national parks were carved out a hundred years ago, MPAs are an evolving tool. And we're learning that what works on land isn't always a good fit in the sea. For example, because the sea is a three-dimensional place with fewer physical barriers than the land, marine organisms tend to travel, either as migrating adults or drifting larvae. Research is showing that we need to allow safe passage from one protected area to the next. An even better way to safeguard the wild sea than setting up individual MPAs may be setting up networks of MPAs. California has embraced this idea, creating a network of 124 MPAs that includes 16 percent of the state waters and is designed to encourage connections between ecosystems. Getting the network off the ground required buy-in from scientists, fishermen, and state agencies. It wasn't easy work. The effort to establish the MPA network failed twice before it finally succeeded.

Besides understanding the seas better and expanding our networks of marine protected areas, there are other ideas out there about preserving our wild oceans. In the same way we try to protect large endangered animals, such as elephants and rhinos, we can try to protect individual members of the ocean's ecosystem. A good place to start would be with the big fish, like bluefin tuna, swordfish, marlin, and sharks, all of which have declined in number by 90 percent since 1950. But fish at the lower end of the food chain need some protection too. About 40 percent of the fish taken from the oceans in 2011 were small fish that school in huge swarms, making them an easy mark for industrial fishing vessels. These small fish are the foundation of the ocean's food web. If they are overfished, larger fish have no chance.

The good news is, we haven't gotten to that point yet. In the ocean, while the wild spaces may be diminishing, they still exist. There are still spaces that are vibrant and healthy, where the water's not a farmyard, where wild animals still swim through the ocean's currents.

That's what Lucas reminded me as we walked down the trail. When it comes to the ocean, "we still have a choice."

Gleaming white and aerodynamic, Japanese bullet trains going other places blasted by Lucas and me like vehicles coming at us from the future. We were waiting on the train platform for a Shinkansen train to take us from Hiroshima to Fukuoka, which is on Kyushu, the southernmost of Japan's four main islands. Despite warnings that we'd find no giant jellies and that we'd be unable to communicate in the remote parts where we'd have the best shot at seeing them, our plan was to get as far out into Sea of Japan as we could in hopes of spotting a giant jellyfish either from a beach or a ferry.

A day earlier, after much phoning and with translation help from a very patient concierge at Lucas's hotel, we discovered that on Tsushima Island, where giant jellyfish had been reported, all the lodging was booked. Frustrated but not daunted, we secured rooms on a nearby island called Iki, figuring we'd get as close as we could. A few hours later, Lucas received good news from Shin by e-mail. He forwarded the note to me:

Hi Lucas:

I called my friend who is a set-net fisherman. He is acting chairperson of Tsushima city council and also runs a small inn. The rooms are not available today, but are available on Sunday. This weekend, there are some meetings in Tsushima, and hence many hotels are full.

Below, you will find the itinerary to Tsushima.

Today (10/20): Stay in a hotel in Hakata (Fukuoka).

Tomorrow (10/21): Take a ferry from Hakata port to Izuhara (Tsushima). From the ferry deck, you will be able to sight *Nemopilema* [giant jellyfish] near the sea surface. Ferry time table: 10:00 (Hakata port) 14:40 (Izuhara, Tsushima).

From Izuhara port to the fisherman's inn, his daughter-in-law will pick you up at the port (this is tentative, I will confirm this with the fisherman to email you), or you can take a taxi to his inn (a bit far and costs you much, >10,000 JPY). Dinner and breakfast in his inn are mainly local seafood. If Juli has a problem with fish, please let me know.

Monday (10/22): After the breakfast, he will take you to his set-nets. I hope you will see some *Nemopilema* in there. You will return to the inn, and take a taxi (you can ask him call a taxi) to Izuhara port.

Time table of high-speed boat: 13:15 (Izuhara) 15:30 (Hakata port). I will meet your boat around 15:30. If you will not be able to catch this boat, use the late ferry (1500 Izuhara 1950 Fukuoka).

Cheers, Shin

"Yes!" This was a plan, a real one. Details and reservations included. We had a shot at seeing giant jellyfish. I thought of the single Japanese phrase I knew, compliments of the old Styx song. *Dōmo arigatō*, Shin. This was a gift.

Our Shinkansen slowed to a stop in front of us. Its doors slid effortlessly open. Lucas and I boarded and settled into our comfortable seats. We released our seat backs, put our computers on the tray tables, and connected to the—oh, yes—free Wi-Fi. The snack lady was already pushing her cart up the aisle, and I bought the hot, sweet coffee in a can and some sort of salty, crunchy treat. I had grown to love the canned coffee found in vending machines all over Japan. But here on the bullet train, rather than a clunk into the retrieval receptacle, I received my canned coffee with a bow from the attendant.

"So, let me ask you something weird," I said to Lucas. "We talked about this a bit when we had a phone conversation about a year ago. You said that polyps are medusa factories. And a single polyp can live for many years and produce many seasons of medusae. So do you think it makes sense to think about the polyp like a tree and the medusa like a flower or a fruit? The reason a tree produces fruit is to move its offspring to another location. Isn't the medusa like that? It's a way for the polyp to move its genes somewhere else."

Lucas didn't quite buy into my theory. "The medusae produce gametes that then make some larvae that then settle. It's not exactly the same as a tree and fruit."

"Yeah, but a fruit produces a seed that then plants itself in the soil. I think we think about the jellyfish all wrong, that they are closer to plants than animals."

He shook his head. "I've heard that idea before, but it doesn't sit right with me. Jellyfish are animals. They move with muscles. They prey on other animals. You lose a lot of the complexity of what jellyfish are doing if you think about them like plants."

"Okay, but you know how sometimes when conditions get bad a plant will just fruit like crazy? It's trying to get its genes to a place where conditions are better. Couldn't that be what's going on with jellyfish in our world today? Conditions are bad, so polyps are all fruiting like crazy?"

"You need to be really careful about that kind of idea. I don't think we know enough about what's driving jellyfish populations to make that kind of connection."

"All right." I still held on to my medusa-as-fruit idea but recognized that the logic was flawed. I decided to change the subject. "So, what do you really think about jellyfish abundances? Do you think they are increasing globally?"

He said, "It's still a question. The paper by the NCEAS group that addresses whether jellyfish are increasing globally is in press now."

I was really hoping to talk about this paper with Lucas, because I had heard that there had been a serious disagreement over the interpretation of the data collected by NCEAS. There'd been so much wrangling that the authors held a special meeting in Santa Barbara to work through the arguments face-to-face. The problem had its roots in the historical disregard of jellyfish and the inconsistent data collection. The NCEAS team collected thirty-seven datasets spanning the years 1940 to 2011. Before 1970 there were fewer than ten datasets published on jellyfish in any year. After that, the number of datasets increased into the twenties and thirties. Some NCEAS members believed that comparing the period before 1970 with the period after 1970 didn't make sense. During the years of sparse data, each bit of information carried more weight than during years that were more data rich. This could skew the analysis by giving a disproportionate impact to earlier data relative to later data. Other members argued that if data existed, it needed to be included, otherwise deciding what data to include and what to exclude imposed biases. It would be cherry-picking. These members pointed out that there were statistical methods for dealing with the changes in the size of the databases. There was merit to both of these arguments.

Lucas explained that if all the data were included, the analysis showed that the abundances of jellyfish oscillate in a cycle that repeats roughly every twenty years. The most recent upswing started in 2004, and we're still in it. Jellyfish have been noticed more, not because of some aberration, but because we are on the part of the normal cycle that's tracking upward.

But if you did the analysis excluding the sparse data before 1970, the conclusion was different. Over the past forty years, the data revealed an oscillation, but that up-and-down cycle was superimposed on an overall increase in jellyfish abundances.

This difference in how to perform the analysis—with or without the data before 1970—caused the rift in the NCEAS group. The majority eventually came down on the side of including the entire dataset and published the paper as "Recurrent Jellyfish Blooms Are a Consequence of Global Oscillations." In the end, Lucas decided to pull his name from the paper because he didn't think the pre-1970 data should be included.

When the paper was published, my Google alerts rang with headlines like these: "Jellyfish Blooms Wax and Wane in Normal Cycles," "Researchers Disprove Reported Jellyfish Increase," and "Why the Predicted Jellyfish Bloom May Be a Bust."

The scientists who wanted the entire dataset included in the analysis see these sorts of lukewarm headlines as more honest. In 2016, some of the same members of the NCEAS group released a paper investigating how the headlines that had helped fuel my interest in the story of jellyfish, such as "Meet Your New Jellyfish Overlords," made their way into not just the mainstream media, but also the scientific literature. A 2001 paper was published with the title "Jellyfish Blooms: Are Populations Increasing Globally in Response to Changing Ocean Conditions?" While the author—one of the most respected in the field—couldn't answer the question because there wasn't enough data to do so, just raising it seems to have set off a bias toward the answer yes. The following decade was filled with speculation about changes in jellyfish numbers. Jenny Purcell published a paper titled "Jellyfish and Ctenophore Blooms Coincide with Human Proliferation and Environmental Perturbations." Another paper, "The Jellyfish Joyride: Causes, Consequences, and Management Responses to a More

Gelatinous Future," showed how human impacts like overfishing, climate change, and coastal development could change the ocean to favor the growth of jellyfish populations. All that speculation hardened into claims that jellyfish were taking over the oceans.

It wasn't until Lucas and his colleagues published their study in 2012 that someone actually did a global analysis. They discovered that in 60 percent of the large marine ecosystems, large areas of the ocean near the coasts that are ecologically cohesive, jellyfish numbers were up. But the paper was criticized for using anecdotal evidence from fishermen and news reports. Today, the scientific community remains split about whether jellyfish numbers are increasing globally or not. One side continues to say they are, while the other continues to say it's an oscillating system. What both sides agree on is that more data and more research are needed to understand what jellyfish are doing out there.

"But really," Lucas said as our train neared our destination, "the question of whether or not jellyfish are increasing globally is not as important as the other questions you can ask about jellyfish and the environment, questions like, What's happening in a specific location to change jellyfish abundances? Why are those things happening? Should something be done to mitigate those problems at a local level? And if so, what?"

Our conversation made clear that the question of whether jellyfish numbers are increasing or decreasing might really just be the stuff of headlines. Much as we might be attracted to a clamoring sound bite, it's got limited value. What really matters is the long-term wrangling over the question and the search for answers. In our increasingly polarized society, we can take a tip from the NCEAS scientists, who continued to hash over data, to publish papers, and to collaborate despite their disagreement. Progress occurs when opposite sides engage with each other rather than talk past each other. Debate and disagreement done right force us to find new ways to answer questions, to look for mistakes, to reevaluate how we understand the world that we share.

Stalking the Beast

Lucas and I walked down the gangway amid the bowing of the crew on either side of us. We had just disembarked from a ferry on the small island of Tsushima, which is as far from the Japanese mainland as you can be without being in Korea. A small white car swerved into a parking spot at the ferry landing, and a man wearing a white bandana around his head and knee-high white plastic boots jumped out and bounded toward us. It was easy for him to recognize us, the only foreigners on the island. Lucas especially, at six feet tall with a shiny bald head, stood out from the crowd. The bandana-wearing, booted man kinetically motioned us to his vehicle and we climbed in.

We never found out who the man in the white boots was, but he treated us to the best tempura shrimp I've ever eaten. Then he passed us off to a woman who took us on a long and twisty drive toward the northern tip of the island. As she drove us over the narrow island roads carved into dark igneous rock, we settled into the frustrating quiet of people who would like to speak to each other but can't. I stared out the window at the dense trees, thick with foliage. Here and there, a maple turning shocking autumn red interrupted green. It was the time of the rice harvest, and the smell of burning brush filled the air. Bundles of golden stalks hung drying from racks made of bamboo that stretched from end to end of each rice field.

Farmers tended fires that burned the cutoff bases of the stalks, returning their nutrients back into the ground. We passed over a tidal river. I saw a woman hand-watering her garden with a ladle. A scarecrow made from a worn-out mannequin presided over her crops. I felt very far from home.

Eventually, we pulled up in front of a fairly large house that sat on an inlet, a natural harbor surrounded by a network of docks and embankments. Our hostess, Teruko Sakumoto, graciously invited us inside. Before we entered, we traded our shoes for the house slippers she provided. She escorted us up a polished wooden staircase to the second floor. Lucas's room was to the right, mine to the left. In the hallway were two bathrooms, one for women and the other for men. The innkeeper pointed to slippers outside the bathroom. We understood that we were to change out of our house slippers into bathroom slippers before we entered.

I peeked into my room and saw a lot of open space, nothing like the clutter of the bedrooms at my house in Austin. The floor was covered with finely woven straw mats edged in green silk. Each was roughly the size of an unrolled sleeping bag. These were tatami mats, traditional Japanese flooring. They were placed together like a perfect jigsaw puzzle so that all the junctions were T-shaped rather than +-shaped. I had read that such intersections were considered inauspicious. In one corner was a low table the size of a stool. It held a mirror and a blue box of tissues. I quietly set my bags next to it. Across the room, there was a TV on a stand and a pile of bedding folded neatly. A sliding door opened onto a balcony overlooking the harbor. Closet doors embellished with mountains and billowing clouds filled the right side of the room.

But now was not the time for relaxing. We were here to find giant jellyfish. Not wanting to invade the uncluttered quiet in the room, I unpacked just enough to find my camera. Downstairs, I saw Lucas making little walking motions across one hand with the fingers of his other hand to indicate to our hostess that we were headed out. We changed back into our street shoes and strolled past fishing boats and docks. We scrambled over piers and nets, and past a sign written in Japanese and Korean explaining the history of this tiny port, and searched as deep under the water surface as we could manage for the shadow of something big and gelatinous. We saw lots of debris and trash but no sign of giant jellyfish.

As we returned to the inn, a car pulled up and a man emerged. This was our captain, Yoshifumi Sakumoto, whom we called Sakumoto-san. He wore a blue baseball cap splattered with saltwater, and a gray sweatshirt under a blue fishing vest. We shook hands, and then Sakumoto-san pointed to his watch and said, "Seven." He gathered the tips of his fingers and thumb together and put them to his lips, a gesture that Lucas and I took to mean, "Dinner is at seven."

Promptly at seven, Lucas and I descended the wooden staircase and entered the dining room. The same clouds and mountains that floated across the walls of my bedroom graced the dining room too. Lucas and I settled ourselves on pillows on the floor. The low wooden table was set with delicate blue-and-white china. Chopsticks rested on a large plate in front of me. A glass sat on a saucer. Beyond my place setting was a bowl of ceviche, a platter of the largest, most luscious sushi I'd ever seen, a serving bowl laden with sashimi and garnished with sprigs from the garden, and two huge roasted squid. And that was just for me. Lucas had an identical portion on his side of the table. It was an enormous bounty.

"Are we supposed to eat all this?" I asked.

"I think it would be very rude if we didn't try our best," Lucas replied.

Just then, Sakumoto-san entered the room. He carried a steaming cauldron of chicken soup and a tub of cabbage and vegetables. He set the soup on a warmer to my left, switching its dial to heat. Using chopsticks, he slowly added vegetables to the chicken soup, stirring them gently. Then he left us. The quantity of food was ridiculous.

"What do we do now?" I asked. Lucas poured us each a glass of beer, really the only sensible response. Sakumoto-san reappeared. He checked on the vegetables cooking in the soup.

Lucas said, "This is too much!" Lucas made a wide gesture with his hands, trying to pantomime "too much."

Sakumoto-san smiled benignly and stirred the soup with a chopstick. He acted as if he didn't understand us, but he did. Sakumoto-san's wife, Teruko, joined us. She was holding two more large platters of food, containing an assortment of fish fritters, as well as grilled conch.

Sakumoto-san pretended to get an idea. He called Shin, had a short conversation, then passed the phone to Lucas. "Shin, please tell Sakumoto-san and his wife we are so grateful for how well we are being treated. Please tell them there's enough food here for a banquet. We can't possibly eat it all." Even as Lucas was saying this, Sakumoto-san was ladling out bowls of chicken soup, holding back a smile. It was a game.

After Lucas got off the phone with Shin, our hosts left us to contemplate the outrageous amount of food. Lucas took a bite of a fish fritter. "You know what Shin said? That just as wonderful as we are being treated, it's as wonderful for Sakumoto-san and his wife to have us here. Westerners never come to really rural places like this. He said, even though we can't communicate with them, we are all human."

We are all human. I considered it. All people enjoy the same things: food, laughter, the excitement of meeting people from the other side of the world, the fun of playing a joke on them. The abundance and the warmth of the dining room unlocked the philosophical side of me. "You know, maybe that's where the hope lies. The recognition that we are all human and that we are all stuck here on this planet together. Maybe that fact will be what eventually gets us to clean up our act." The idea seemed so clear to me seated on the edge of a sea where giant jellies swam ever more frequently. Every beach I'd seen since I got to Japan was a mess. On the ferry ride, the water was filled with trash. Those were only the problems I could see, but it was obvious the ocean was being treated like a dump. But at home in Texas, where I'm surrounded by land, not to mention the power of the petroleum industry, the urgency slips away. Plus, I get busy. Removed from my everyday experience, the clarity that things are very wrong becomes blurred.

Sakumoto-san returned, interrupting my musing. He pointed to his watch, "Six. *Echizen kurage*." Then he pointed upstairs to our rooms. The jellyfishing boat would leave at six a.m. Time for bed. We were improving at this game of pantomime.

I awoke at four in the morning, crawled out of the bedding, and dressed in my grungy fishing clothes: old yoga pants, old T-shirt, old jacket, tennis shoes. I checked and rechecked the batteries and seals on my new underwater camera, pulled out my snorkel and mask, and put everything in an old backpack. I didn't know what this trip would entail, but I hoped for the chance to get in the water with a giant jelly, to watch its majestic bell pulse and wave. Around five a.m., I heard a rooster groggily half crowing, so I headed downstairs. Sakumoto-san was already up and ready, wearing blue rain gear head to toe.

We walked outside and crossed the street to the dock. Sakumoto-san's boat was waiting for us there, a long low vessel with a small wheelhouse. Three crew members were already preparing for the day's fishing, moving hoses and plastic tubs and baskets to their proper locations on the deck. Like Sakumoto-san, they were dressed in full rain gear. The men worked together like a well-oiled machine. Few words were exchanged among them.

Sakumoto-san motored across the glassy bay and around the outer coast of the island to the south. As we approached the floats that hold up the top of his nets, one fisherman used a hooked pole to grab a line and attach it to a winch. A complicated ballet of winching and unwinching followed, until the net was close enough to be attached to a crane in the center of the boat.

Sakumoto-san pulled a lever, extending a crane that lifted the net high in the air. As it rose, a fisherman continuously hosed it down with a blast of seawater. At one point, Sakumoto-san stopped the crane, and two fishermen unzipped a giant zipper that was sewn into the net. Then the crane lifted the center of the net higher as, hand over hand, two fishermen brought the lower sections of net onto the deck of the boat. There was no safe way to swim amid this heavy equipment. I wouldn't be using my underwater camera or my mask and snorkel. If we did find a giant jellyfish, I wouldn't be seeing it from the water.

Lucas and I stayed out of the way as the fishermen began sorting the fish. Tuna-like jacks went into a well below the deck, small fish into a styrofoam cooler. Puffer fish were tossed back into the water—perhaps they were not yet in season—as were many long needlefish that I guessed were too bony to eat. Squid went into a large yellow basket that was quickly stained black with their ink. Sucking in and then expelling air, the squid were a cacophony of wheezes and snorts. I noticed that one crew member was inked in the face.

After the initial sorting was done, Sakumoto-san beckoned us. He pointed into the net and said, "*Echizen.*" Lucas and I stepped forward and peered down. A massive dark pink blob drifted among the wriggling fish, and then it sank out of sight.

Lucas and I stepped away so that the fishermen could pull out more fish from the nets. They motioned for us to look again. We saw the *echizen*, now tilted to expose its frilly, dark red underparts. It was both beautiful and beastly.

Sakumoto-san said a few words to his crew. They leaned over the edge of the boat and joined together to haul the jellyfish onto the boat's deck. The creature was heavy, awkward, and easily ripped.

I crouched near the jellyfish. I guessed it was three or four feet across. Its mauve bell was nicked and scarred by the nets. I reached out to touch it. It was leathery and cool, wet but not slimy.

"Can you turn it over?" Lucas asked Sakumoto-san, gesturing a flipping motion. Two of the crew who were wearing gloves turned the creature over, exposing the stinging tentacles and mouthparts, a ruddy frill. Now I could see that there were eight crimson oral arms surrounded in a confusion of spaghetti-like tentacles. These huge oral arms were lined with many, many tiny mouthlets, too small to see, each of which opened into the digestive system. The creature was optimized to eat constantly and ingest everything in its path, to grow really big really fast. When they are young, giant jellies pack on 11 percent to their weight each day. By adulthood, the increase declines to 2 percent each day. Still, that would be like me gaining fifteen pounds a week.

As we looked at the jellyfish's undersides, Lucas brushed his hand close to the oral arms. "Ouch," he said, shaking his hand. He'd been stung.

"What's it feel like?" I asked.

"A bee sting. I definitely feel it." A couple of hours later, he told me the discomfort was gone. Giant jellyfish don't have the most venomous sting, but there are reports of fatalities from them. Because they are so large, they can overwhelm a swimmer whose skin isn't protected by a wetsuit.

I sat back on my heels and thought, *Whoa, I'm sitting next to a giant jellyfish.* Despite all the uncertainty of getting to Japan and then to Tsushima, and the sheer unlikelihood of actually seeing this animal, here it was. I was next to a creature that had been born in China and traveled hundreds of miles through the sea, around the Korean Peninsula, only to get stuck in a net off the tiny island of Tsushima. And then to be hauled up out of the depths before the eyes of two people from North America.

Lucas and I high-fived each other, but it felt forced. I didn't feel elation or happiness. I didn't even feel satisfaction. I certainly felt fascination, and I felt that I should do more with this creature than just see it. There was no victory in reaching this goal. Instead, there was a real emptiness to the moment—born, I think, of the realization that the massive animal would die before me.

The fishermen moved on to the next net, performing their complicated choreography with the winches and zippers. This haul came up with somewhat fewer fish and a smaller *echizen*.

Out of water, the two *echizen* deteriorated quickly. Their bells deflated; the tentacles and oral arms broke apart. The degradation was like the melting of the Wicked Witch of the West but for the opposite reason. When immersed in air—an environment it was never meant to experience—the giant jellyfish simply melted away.

We had been out on the water for about four hours, and the fishing was done for the day. Some of the crew reset the nets. Others continued to sort fish, putting them in various containers and baskets. Sakumoto-san backed the boat away from the fishing grounds and aimed us toward the harbor. Once the haul was completely sorted and the baskets were all secured, the crew lit cigarettes and made phone calls. One fisherman hosed off the deck. Black squid ink mixed with the melted gelatin of the *echizen.* It ran through the gaps between the deck and the boat's sides, flowing in rivulets back to the sea.

16

Jellyfish al Dente

The ferry rocked roughly from side to side and from front to back. Its engine cried out when it lifted from the water. Deep ditches and massive dunes formed in the sea surface and then played tackle with one another. Blasts of spray obscured my view out the window, but I didn't really want to look anyway. I closed my eyes and saw newspaper headlines about ferryboats sinking. I didn't want to die in a Japanese sea. I plugged my ears with headphones to drown out the *whomp-whomp*ing of the boat against the waves with the comfort-food sounds of Bruce Springsteen, as if somehow the Boss would keep me safe. I gripped the armrests as if I could hold up the boat with my hands. The other passengers shared none of my anxiety. Some were sleeping, others chatting or reading or snacking. They were probably used to the ride. They knew that if the weather were too bad, the ferry company would have canceled the trip.

And they almost did. When Lucas and I arrived at the ferry station, we were told by an interpreter that a very strong weather front was passing through the region. He said that the ferry back to the Japanese mainland may be delayed until conditions improved. About an hour before departure we got word that everything was on schedule. The seas were passable. As I planted my feet firmly on the shifting floor, I really hoped so.

Lucas looked over at my drawn expression and my white knuckles. "I didn't know that boats made you nervous."

"Neither did I," I answered. My fear wasn't something I would have admitted in my grad school days when I was trying to show that I could handle anything, that I had what it took, or what I thought it took, to be a scientist. But now it's easier for me to admit when I'm scared. Age has helped me understand that being scared doesn't make me weak; it just makes me honest.

Of course, the ferry arrived safely at its dock. And my palms stopped sweating immediately after we arrived at the ferry terminal. Right on schedule, Shin pulled up to take Lucas and me to our next stop on what had become our Japanese jellyfishing odyssey. Gone was the formal suit–wearing scientist from the meeting in Hiroshima. In his place was a relaxed researcher in black T-shirt and light-blue sweatpants with a trunk full of giant containers holding seawater. Lucas and I climbed into the car, relaying the details of our *echizen kurage* adventure and thanking Shin for his part in it.

Our destination was a town about an hour away called Yanagawa, which sits on the edge of the Ariake Sea. The Ariake Sea is really a huge bay carved into the island of Kyushu. The sea's very shallow northern section, where we were headed, has the largest tidal fluctuations in Japan, as great as twenty feet. These tides carry great quantities of sediments, which contain loads of nutrients. The combination of sunlit shallow water and nutrients is just what small phytoplankton need to grow well and provide a rich base for the food chain. This northern part of the sea had long been extremely productive, famous for large catches of shrimp, fish, clams, and mussels.

"But things have changed recently. The abundances of fish, bivalves, and shrimp have decreased," Shin explained, launching into the first of what would be several wonderful extemporaneous lectures he'd share with us over the next two days. Shin has the uncanny ability to turn any unassuming space, such as the inside of a compact car, into a lecture hall. He said that the ecosystem was degraded, largely because of pollution, but also due to aquaculture. The Ariake Sea was home to an enormous seaweed-growing operation, one that produced more nori, the flat sheets of seaweed used to wrap sushi rolls, than any other place in Japan. About thirty years ago, farmers started using organic acids like citric acid to kill disease-causing organisms that grow on the seaweed. According to Shin, this increased the production in the Ariake Sea from about 3 billion dried sheets of nori to about 5 billion in 2005. The acid is poisonous to plankton, however, and so artisanal farmers who also catch shrimp, crab, and bivalves, all animals that eat plankton, complain that their harvests suffer as a result of the acid's use in seaweed farming.

But it's not quite that simple, Shin cautioned. Some suspect that the phosphorus in the acids may be the real culprit. Phosphorus is a fertilizer that can increase the growth of phytoplankton. When phytoplankton grow, they ooze out some of the sugars that they make during photosynthesis. Bacteria love sugar, so they grow like wild in the presence of a phytoplankton bloom. Bacteria use up the oxygen as they grow. Thus increased phosphorus results in decreased oxygen, suffocating animals that are unable to swim, like bivalves in an aquaculture farm. And that may not be the whole story either. The decreased catch in the Ariake Sea could also be associated with a landfill project in Isahaya Bay, an inlet on the western side of the sea, which could be causing pollution and disrupting normal coastal ecosystems. Some suspected a dam that was built on the largest river flowing into the sea was the problem. Or the decreased catch could have been a result of something else entirely.

"So the harvest has declined," Shin concluded. "There has been an ecosystem shift. The formerly high productivity and biodiversity of this sea have diminished. The Ariake Sea is now mainly a seaweed culture field in the cold seasons and jellyfish-dominated sea in the warm seasons." This year the red jellyfish, *Rhopilema esculentum*, or in Japanese *bizen kurage*, had been the benefactor of these changes. "The *Rhopilema* abundance is increasing more and more. And this year, there is a dramatic outbreak of the natural population. The fishermen used to catch a full boat of seafood, but no longer. So today, they are jellyfishing."

About twelve years earlier, one of Shin's students had introduced him to a fisherman named Katsueda Aramaki, who had fished the Ariake Sea for years. Shin now visited regularly, using the fisherman's boat as a survey platform to study jellyfish. Shin said, "Aramaki-san helped keep a record of the jellyfish sizes. In May, the *Rhopilema* would be the size of a softball. By June, they would be the size of a basketball, and by July, a beach ball. In August, they stop growing, and then they die in December." It was October, so the jellyfish, if we got to see them, would be at their largest, just before senescence.

Near the town, Shin exited the highway, then soon pulled off the narrow road and drove down the fisherman's driveway. "I think we should make a stop at Aramaki-san's house before we go to the hotel. A front is supposed to pass through, so we should make sure that the weather will be okay for us to go fishing tomorrow." It had just started drizzling. I wondered whether this was the same front that had made for such a rough ride back from Tsushima.

A skinny white dog stood up from a nap on a pile of old fishing nets and barked loudly to warn the family of our arrival. Fields stretched out behind the dog's territory. Like Tsushima, it was harvest time here, and a rice field smoldered nearby. We scurried through the drizzle and under the cover of the garage. A man in his thirties greeted us. He was Teruyuki Aramaki, the fisherman's son. He started his fishing career working the seabed for clams, but the decreasing harvest forced him to abandon that business. Now he was a jellyfishermen.

One side of the garage was set up to process small quantities of jellyfish. White foam coolers and plastic tubs and buckets in various pastel shades sat on the concrete floor. Each was filled with saltwater and jellyfish parts. Large bags of salt and alum, maybe fifty pounds each, were stacked along one wall. Shin reached into a blue tub and pulled out half of a jellyfish bell from its saltwater bath. It was so heavy that he had to hold it in two hands. He spread the pickled jellyfish so that the underside faced Lucas and me. It was as wide as his body and draped downward about a foot and a half. It looked like the underside of a gigantic mushroom, ribbed in the parallel lines of dark red muscle. Shin said, "For export to China they clean the body, remove the gonads and tentacles, and get rid of the umbrella. This umbrella I'm holding is considered the cheap part of the jellyfish, and it would be discarded. But the fisherman's family eats it because they don't want to waste it."

"Why don't you eat the bell?" I asked. The cheap cut seemed to hold so much meat.

Shin explained that if the bell gets too thick, it's hard to remove enough water to toughen it up. "The thickness matters." Unlike the cannonball jellyfish that's fished in Georgia, the oral arms are the most prized part of the *Rhopilema* because they are the right thickness to acquire the sought-after crunch. "You soak the jellyfish for three days in the salt and alum bath." He pointed to a green bucket. The saltwater in which the jellyfish soaked was murky and foamy. "Then you take it out, and put it in fresh saltwater." The example of this second soaking was in a blue bucket. The water looked clearer, the jellyfish firmer. "And then a third soaking for three more days. After this it is crunchy." He gestured to a yellow bucket. The water was less murky, and the jellyfish looked opaque and solid.

Shin pulled out two pieces from this last bucket. They looked like thick red kale, bumpy and frilled on the edges. "At this point you could bring the jellyfish to auction and export it. If you soak it for too long you can remove too much of the water and destroy the quality."

I asked Shin something I'd been wondering about: "Can you eat the giant jellyfish?"

"All jellies are at least edible. But most people consider the crunch the important part, and these red and white jellyfish can be processed to produce that crunch. The *echizen* are too watery." I recalled the slurry of melting *echizen* and squid ink on the deck of the fishing boat in the morning. In the years since, some people have begun to develop methods for curing giant jellyfish, hoping to make a product from the prodigious biomass of the animal. Shin continued: "A chef from Oita prefecture not far from here developed a method for processing moon jellies. They have so many there. Some housewives process it for family consumption, and some people do enjoy the taste. I think that chef even has classes on preparing it still today. I tried the prepared moon jellies, but they're not that tasty." Oh yes, my pets Peanut, Butter, and Jelly could have actually been a meal.

The younger fisherman invited us into his home to greet his father, Katsueda Aramaki, and presumably to discuss the weather. When we entered the house, the family was sitting on the floor around a low table, finishing dinner. A large television rested to one side of the table, broadcasting a baseball game. Space was made for Lucas, Shin, and me at the table, and we sat on the floor with the TV to our backs.

The senior Aramaki-san sat at the head of the table in a legless chair that allowed him to lean back. He gave off the air of a patriarch, the way my grandfather used to preside over our family dinners. There were two women at the table; I assumed that they were Aramaki-san's wife and daughter-in-law. The younger woman rose from the table and returned with hot green tea for us. She peeled and sliced apples and Asian pears, and served them to us with toothpicks. Lucas and I sipped and munched and listened to the Japanese swirling around us.

Shin discussed the weather with the father and son. He translated for us. A strong front was going to make jellyfishing tomorrow difficult. We would have to wait until afternoon to see if conditions improved. As I'd learned from Wynn Gale and cannonballs in Georgia, *Rhopilema* also have good control over their vertical position in the water and dive if the weather is too rough. After the bouncy ferry ride earlier that day, I was not too eager to get back on another rough boat anyway.

Through Shin, Lucas and I described our adventure with Sakumoto-san's fishing nets. I pulled out my phone, scrolled to a photo of me crouching next to the giant jellyfish, and passed it around the table. Shin took a look and said something in Japanese. Aramaki-san agreed.

"What is it?" I asked.

"Well, that's not much of an *echizen*."

"Really?" It seemed pretty big when the crew was pulling it out of the water. Lucas and I had estimated that it weighed about 250 pounds.

Shin gestured to indicate the length of the table. "I've seen them as big as this." It was a six-foot table; jellyfish that size would dwarf the one we saw. Okay, so ours was puny.

The daughter-in-law said something to Shin, and he translated. "She would like to serve some jellyfish. Usually they have some on hand, but they don't right now. So she will go get some that they have just finished processing and rinse it for us. She's sorry, but the jellyfish will still be kind of salty. It's supposed to be flavorless. Jellyfish is all about texture." I watched her in the kitchen, putting the jellyfish in a colander and running water over it, just as the woman in the Asian market had counseled me: Wash it many times.

The door opened, and a teenage girl entered. She was wearing a volleyball uniform and carrying a gym bag. She moved like an athlete. Shin introduced her as the younger fisherman's daughter. "She won her game, but now she has to do some homework." She smiled and waved. Then she was off to her room, just like my kids at home. The fisherman's daughter-in-law returned and placed a small white plate of thinly sliced jellyfish in front of Lucas and me. The pieces were the reddish color of boiled shrimp and they had a translucent quality. Next to the jellyfish, she placed a small bowl containing a mixture of soy sauce and vinegar. She handed Lucas and me each a set of chopsticks. I was so curious to compare them with my jellyfish salad. And when was I going to get the chance to try jellyfish prepared by a jellyfisherman's family again?

I picked up a slice. I didn't dip it in the sauce at first, wanting to taste it unadorned. I chewed. It was not noodle-ish like the jellyfish salad I made in Austin. It was a much crunchier, fresher-tasting version—like fresh pasta compared with boxed. I caught the eye of the senior Aramaki-san. He tapped his fingernail to his teeth. Yes, he was right, it was al dente.

There was a break in the rain as Lucas, Shin, and I walked from our hotel to a nearby street where several restaurants are located. We decided on an Italian place, and Lucas and I turned the ordering over to Shin.

When the antipasti arrived, Shin asked, "Juli, remind me, where are you from?"

"Texas. Austin, Texas."

He said, "I was in Texas once. I was in Port Aransas, where there's a marine lab." I nodded. I knew that lab. It's affiliated with the University of Texas at Austin, and I had gone there soon after I moved to Austin.

Shin continued his story. "I was visiting a scientist, who took me out to dinner at a local restaurant. On the wall, there's a Confederate flag and a Texas flag. And I remember looking at the two flags. The scientist noticed me staring. He said, 'The United States is one thing, but Texas is Texas.'"

And I agreed, as a transplant, that Texas is Texas. For better or worse, Texas is its own place, with its own customs, behaviors, and motives. People wear cowboy boots, cowboy shirts, and cowboy hats—and not just as a costume. They go two-stepping to twangy guitar music. They cook barbecue and eat pie—a lot. There's a rumor that it's illegal to pick bluebonnets, the state flower, from the side of the road; it isn't true, but all Texas children repeat and obey the rule. A lot of these practices are charming and enrich the lives of Texans, but others are just plain confusing. As an outsider, I find it hard to understand why, for example, you would hang on the wall a flag that represents a war lost a century and a half ago that was fought to defend slavery.

Shin asked, "So Juli, what made you want to write about jellyfish?"

"I know it seems strange. Initially it seemed strange to me, too," I said. I explained that I'd been fascinated by invertebrates since I first went diving in Israel but that my life had taken me away from the sea. I told him how writing captions for an article about ocean acidification made me wonder whether jellyfish would really be winners in a future acidified world, and then I fell down the rabbit hole when I started exploring all the amazing characteristics of jellyfish.

I told Shin that there was something personal too. In the past, my writing had been relatively prescribed—state standards and editors steered my work. But this was a project in which I could express my own thoughts and passion. Jellyfish had become a better-late-than-never vehicle for me to explore the threats to the ocean's future. They're a way to start a conversation about things that can seem boring and abstract—acidification, warming, overfishing, and coastal development—but that are changing our oceans in fundamental ways. On the train, Lucas had pointed out that whether jellyfish numbers were on the rise globally was the wrong question, as misguided as viewing the diverse ecosystems on an entire continent—the wooded mountaintops, the alpine rivers, the sweeping prairies, the arid deserts, the rich estuaries, the rocky chaparral—as a single biome. Like any particular neighborhood, each ecosystem in the ocean has its own unique characteristics, with distinctive vulnerabilities, threats, and resiliencies. Understanding whether jellyfish will flourish or fail requires attention to those differences and a strong understanding of the ways they interact with other ecosystems. The ocean is a vast and complicated place, and we barely understand how our actions on land affect it. The insight also sheds light on that very first strange conversation I'd had with Monty Graham. I remembered Monty leaning back in his chair in his office at Dauphin Island Sea Lab and giving me his opinion of the importance of global changes in jellyfish abundances: "So fucking what?" It didn't make sense then, but I thought that I'd learned what he meant. Understanding jellyfish means recognizing the oceans in their complexity rather than homogenizing their problems.

"These things are all important," Shin said in his characteristically sage manner, "but it's very hard to determine the nature of jellyfish. Just when you think you understand them, when all indicators suggest that you're going to have a year with a big jellyfish bloom, they disappear. Like this year for the giant jellyfish."

No kidding, I thought.

"And when you don't expect it, jellyfish bloom, like this year for *Rhopilema.* It's hard for scientists like me to learn more about jellyfish, because when we have a lot of funding to study jellyfish, they just don't show up. And when we aren't prepared, their population blooms." Shin took a sip of beer and grinned. "I think they're teasing us."

We toasted to that. Much as we might want to put things in a box that we recognize and understand, some things—maybe the best things—defy categorization.

Jellyfishing

Maybe I was still a little jet-lagged, or maybe it was that my head was buzzing with the jellyfish adventures from the day before, but I was not at all tired when my alarm went off at four-thirty a.m. The storm front was upon us; our plan was to check with Aramaki-san in the afternoon to see if it would be calm enough to go jellyfishing. Until then, we had time, and Shin wanted to check out the Yanagawa fish market. We had to get up so early to make the start of the auction.

The Yanagawa fish market is one of three major markets in the region, and not even close to the largest in Japan. That honor goes to the Tsukiji market in Tokyo, an exchange made famous by photos of rows and rows of beheaded and betailed torpedo-shaped tuna. There, two thousand tons of marine products change hands each day. But whereas the Tsukiji fish market became so popular as a tourist attraction that visitors were essentially banned from the wholesale sections, the Yanagawa fish market was open to all who are willing to get up early enough to see it.

It was raining lightly and still dark when we pulled up, but the market was wide awake. The concrete floor of the open-air pavilion was covered in rows of styrofoam coolers, each containing a bid lot. As we approached, a very loud bell sounded, signaling the start of trading. An auctioneer hightailed it to the end of a row and blew a whistle several times, very loudly.

Buyers wearing baseball caps with plastic ID cards bearing their registration numbers quickly surrounded him. A cacophony of words erupted from the auctioneer. The bidders nodded and raised their hands. The auctioneer paused, made a note, and the winner dropped a piece of preprinted paper bearing his number on the lot, claiming it as his. As a unit, everyone took one step together to the next cooler. The bidding and selling went so fast at times that the whole crowd was slowly strolling down the row together.

It was hard to know who the winner was when the lot was sold. I kept guessing wrong, anyway. I'd be sure that the lot was sold to a guy in a red jacket with 2454 pinned to his hat, but then I'd see a man in a gray sweater, number 52, drop a piece of paper on the lot. Once a lot was sold, the buyer used a short stick with a hook like a gaff to impale the Styrofoam container and drag away his purchase. New sellers used the same hooked sticks to bring their own coolers to replace coolers that were removed after a sale, rebuilding the row. The stick was also useful for tapping the flanks of a fish to determine its quality.

Soon other whistles sounded from other parts of the warehouse. Different auctioneers started their own cluster of bidding, selling, and strolling. The controlled chaos and the noise built together. I could hardly stay out of the path of carts rolling away bigger purchases and of the coolers being dragged to and from rows that continually formed and reformed. There was a palpable but civil excitement.

The array of sea life for sale was like nothing I'd seen before. I recognized the sushi standards: tuna, yellowtail, snapper, loads of mackerel. I saw rays, pufferfish, sardines or maybe anchovies, flatfish, and one very big and very ugly frogfish. There were narrow silver fish I'd never seen before, so long that their whiplike tails flowed well beyond the confines of the cooler. There were thin midwater fish bedazzled with bioluminescent spots on their sides. In one cooler, a cluster of butterfish—known to feast heavily on jellyfish—lay on a bed of ice. Crabs, shrimp, snails, oysters, abalone, two types of octopus, squid, sea anemones (apparently they are eaten fried), green net bags holding understandably aggressive turtles, and many more products of the sea that I was unable to identify were all displayed in wooden boxes. There was even a bit of produce: oranges, some water-plant seed that was in season, and mushrooms. A few sellers packaged their goods into little sushi lunches like those you buy in an upscale grocery store.

Lucas and Shin were motioning to me. I scooted around the rows of coolers and dodged buyers hauling away their purchases to reach them. They were standing in front of a row of wooden crates filled with jellyfish, many red and a few white. A slip of yellow paper lay on top of each one; they'd already been sold.

By seven a.m. the energy died down. The din lessened. The crowd thinned out. Shop owners, restaurant owners, and distributers had made off with their purchases. The few remaining people were loading their goods into the backs of trucks. It had been a two-hour frenzy of buying and selling, shuffling the sea's harvest from producers to consumers.

"It's incredible that so much can be harvested," I said to Lucas.

"It's even more so when you consider that this happens every day," he replied.

I had just watched an immense amount of organic material moving out of the sea and onto our tables. In my head, I tried to grow to a global scale this little market that served just a part of this small sea, but I found I couldn't conceive of it. All I could think was, *This isn't sustainable.*

By the early afternoon, the weather had started to improve, but it still wasn't calm enough to go jellyfishing. To kill some time, Shin suggested we visit the town center of Yanagawa. We walked along a meandering canal, watching tourists float by in shallow boats called *donko bune.* Boatmen pushed these skiffs using long bamboo poles. But for the straw hat instead of the red cummerbund, they reminded me of gondoliers on a Venice canal.

One reason for the visit to town was to check the prices of jellyfish at several fishmongers. The same wooden crates of jellyfish that were auctioned at the market a few hours earlier were now out for display. Red jellyfish, the more common variety, sold for about $4.75 per pound, while the white cost more than double that.

Jellyfish have been neglected not just by science, but by economics as well. Hard numbers on the worldwide catch of jellyfish are hard to come by. Although at least eighteen countries catch jellyfish and a dozen export them, most don't explicitly report their catch. If they do, they categorize them as "miscellaneous" or "marine invertebrates." After our trip to Japan, Lucas pulled together what was known and what was unknown. The United Nations clearinghouse for fisheries information, the Food and Agriculture Organization (FAO), reported an annual jellyfish catch of 350,000 tons. But poring over catch records country by country, Lucas determined that the jellyfish catch actually exceeds a million tons a year and that it has grown rapidly since the 1980s. In 2013 alone, the world's largest jellyfishery, China's, reported catching 211,000 tons of jellyfish. Yet Lucas's research showed the actual number to be 764,000 tons. Both Thailand and India produce over 100,000 tons per year. Smaller fisheries exist in Indonesia, Vietnam, Mexico, Japan, Malaysia, the Middle East, and of course in the southeastern United States. Lucas's estimate didn't include jellyfish that were caught unintentionally in nets, like the giant jellyfish we'd seen on Sakumoto-san's boat the day before. Such discarded bycatch could represent millions of tons of jellyfish hauled out of the seas in addition to the commercial landings of jellies. There might be a lot of jellies in the sea, but we're beginning to pull out a lot as well.

After perusing the shops, Shin suggested we stop at a tourist bureau for a cup of coffee. Inside, hundreds of colored balls hung on strings from the ceiling, each one wrapped in silk string to form intricate geometric patterns. In this region these ornaments are displayed during a Girls' Day Festival. When a baby girl is born, her relatives start sewing the ornaments; they are given to her as a wish that she grows up strong and proud.

As we sipped our coffee, Lucas asked Shin about the *Rhopilema.* "Where do they come from? Are they born in the Ariake Sea, or do they come from somewhere else?"

As this was a tourist bureau, a large map of Fukuoka Prefecture was taped to the back of a wall-size partition. Shin, ever the professor, picked up the partition and turned it so that we could see the map, converting the tourist bureau into a classroom. He pointed to the northern section of the Ariake Sea.

"Here is where we believe the jellyfish come from. We have found ephyra, but haven't found polyps yet." Right, the elusive polyp.

He pointed to all the rivers that flow into the northern part of the Ariake Sea. "The seeding area for all these big jellyfish like *Rhopilema* and the giant jellyfish seems to be associated with rivers and estuaries. Here, in Yanagawa we have a lot of river input." Shin expanded on Lucas's question to include the giant jellyfish, as well. "For the giant jellyfish, the upper reaches of the Yellow Sea in China have a lot of river input. The polyps and ephyra need a lot of nutrients, and the rivers dump that in. We have no Japan-made giant jellyfish right now. All the giant jellyfish that come to Japan are made in China. But that could change in the future. I think that plastic garbage can also be a habitat for the polyps. In the lab, the polyps easily attach to plastic garbage."

It's a real possibility. Recently, researchers working in the center of the North Atlantic collected small pieces of floating plastic from the ocean's surface. When they checked under a microscope, more than half were colonized by jellyfish polyps of the genus *Obelia*. The genus buds tiny, harmless medusae, and it already has a worldwide distribution. But the finding does suggest that we should keep an eye on plastic in the ocean. Besides tricking fish, birds, and whales into thinking it's food and corrupting our food webs, plastic could make jellyfish factories mobile, with the potential to transport invasive species to new places.

Lucas asked, "What about the years when there aren't big blooms? What causes that?"

Shin brought up the ability of jellyfish to form seedlike podocysts. "It's possible that podocysts play a role. Only the polyps of big jellyfish like *Rhopilema* and Nomura's make them, and only when conditions get bad, conditions like low oxygen or warm water. The polyp buds off a podocyst and leaves it behind. Then the polyp moves away from the place where conditions were poor, leaving behind a podocyst that can go dormant for years, enduring the low oxygen, warm water, and low salinity. When conditions improve, the podocyst buds into a normal polyp and then produces medusae."

Lucas wanted to be sure he understood. "So when conditions are poor, rather than producing medusae, these polyps become dormant podocysts. That's why you have years of no blooms. And then when the conditions improve, the podocysts all become active at the same time and produce huge blooms."

"Maybe. It's possible." Shin nodded. "We know that in the lab, the giant jellyfish podocysts can last for six years. We collected some from giant jellyfish polyps in 2006. So far, whenever we've tried to revive them, they have been viable." In other words, when Shin has provided the podocysts the temperatures and salinity that the giant jellyfish polyps thrive in, they spring to life. "It's 2012 and there are still ten podocysts left. So we will find out if they can last even longer."

After the pop-up lecture in the tourist bureau, the sun was shining. We drove back to Aramaki-san's house to see if the weather has improved enough for us to go out jellyfishing. The younger fisherman greeted us with a somber face. Shin relayed the information. "The seas are really rough, perhaps a meter high. It's not good jellyfishing weather. The waves mix up the sediments so it's hard to spot the *Rhopilema*, plus the jellyfish dive when the water is turbulent. But even so, Aramaki-san has agreed to take us out for a brief time. Maybe we'll see something."

With low expectations, we drove a short distance to the boat dock. The vessel was a sloop about forty feet long, with very low deck rails. There was an open cabin made from three walls to protect the driving console. We climbed aboard and headed down a tidal channel lined with many fishing boats. The tall posts that served as moorings reached twenty-five feet above the sea surface to accommodate the great tidal exchange.

At the end of the channel, fields of bamboo and PVC poles stuck out of the water in neat rows, stretching as far as the eye could see. These were the seaweed aquaculture farms, and they bore a striking similarity to the rows of corn that extend for miles in the Midwest. We motored past for nearly half an hour before the farms gave way to the open sea. I understood how billions of sheets of nori were grown here, and how any chemicals used to fend off parasites could have large impacts on the entire sea.

Although the skies had cleared, the conditions were miserable. The wind was blowing hard, and the water was rough and murky. We unsuccessfully scanned the sea surface for the blue bells of jellyfish, all the while jostled roughly in the surge. Aramaki-san pointed to a concrete island, a man-made round structure meant perhaps for storage, in the distance. He gestured that we would use it as a wind shade.

When we pulled behind the island, Shin and Lucas went forward for a better view. They surfed the prow bouncing in the chop. One of them spotted a pale blue blob to the right. By the time I saw it, it had turned a little, showing off a brick-red underside. Aramaki-san grabbed a long bamboo pole with a large net and tried to dip it out, but it dived by the time he got there.

"They're teasing us!" hollered Lucas from the bow, echoing Shin's words from dinner the night before.

Aramaki-san steered the boat back to the lee of the island, and we waited.

Lucas spotted another jellyfish. This time Shin grabbed the net. He dipped it in the water, corralled the creature, and pulled it onto the deck. It was a striking aquamarine and had a much rounder bell than the giant jellyfish, smooth and very solid. When we flipped it, we discovered that the oral arms had been cut out. Some fisherman had already caught this jellyfish, removed the most valuable parts, and tossed it back like a finless shark. Another jellyfish was spotted. This one also had been previously fished and relieved of its oral arms.

We waited a bit longer and caught a third jellyfish, this one containing all its parts, though Shin said it was old and not in great shape. Shin and Lucas brought it aboard. It had a cluster of frilly mouthlets like the giant jellyfish and thick oral arms that ended in red icicles.

Aramaki-san signaled that the fishing was over. He charged back to the dock through the swells, which had grown larger. The ocean spray had us all crouching together behind the pilot's console.

Back at the dock, we put the jellies into buckets so that we could bring them back to the fisherman's house. In the process, I got a handful of oral arm. Brick-colored jellyfish goo full of stinging cells smeared all along my forearm. Aramaki-san motioned for me to wash it off with seawater, which I did, but I was stung a little. I felt an itchiness near my inner elbow, but that was all. These jellyfish were mild.

At Aramaki-san's house, we gathered next to the jellyfish-processing equipment in the garage. Shin prepared to give Lucas and me an anatomy lesson, converting another unimposing space into a teaching laboratory. He had brought along a box of field tools, which we would use to analyze our catch. Here was the answer to the question "What should I do now?" which had confounded me when I watched aimlessly as the giant jellyfish melted before my eyes.

Hanging a shopping basket from a spring scale, we weighed the only complete animal of the several we had collected. Its mass was sixteen kilograms, or about thirty-five pounds. We stretched a tape measure across its bell: sixty-seven centimeters, about two feet across. Shin pointed out the four oral arms, which bifurcated into eight mouthlets and the icicle-like tentacles. He showed us the highly striated muscle along the inside of the bell that allowed the jellyfish to contract forcefully and dive deep when weather conditions were poor. Specialized knobs inside the bell, shaped like drawer pulls, were a characteristic of this species.

Shin slipped on a set of thick magnifying glasses, like those that a jeweler would wear. He looked in the stomachs of the jellyfish to find out what they'd been eating. He reported seeing a lot of copepods, also fish eggs and larval oysters and clams. The jellyfishes' gonads were found alongside the digestive tracts leading to the stomach from the oral arms. Using a pair of scissors, Shin cut out some of this clear tissue and put it in a plastic dish. He started teasing it apart with scissors, looking to see what sex the jellyfish were and whether their eggs and sperm were ripe. After a while, Shin peered up at us. "We caught one female and two males, so we can try to fertilize the eggs." Shin made a slurry of the eggs and sperm in the small plastic dishes, performing in vitro fertilization for the jellyfish. He distributed the slurry among three large containers of filtered seawater that he'd brought from his lab in Hiroshima. In a few days, a student would pipette the surface of the water and place it under a dissection microscope. If this fertilization had been successful, thousands of planula larvae would be swimming near the surface. The student and other members of Shin's lab would try to get the planulae to settle onto plastic plates so that they could study the *Rhopilema* polyps—ones that had never been found in the Ariake Sea.

I dug through my backpack for a plastic bag that I'd been carrying with me for the whole trip. Inside were burnt-orange University of Texas baseball caps. I handed one to Aramaki-san and the other to Shin, and they both graciously received the meager gift. Earlier in the trip, I'd handed similar caps to Sakumoto-san and his wife. It was only a token of thanks for the hospitality, generosity, and warmth that they shared with me.

Shin went into the house to say good-bye to the senior Aramaki-san. His daughter-in-law came to the door wearing a surgical mask. She said her father-in-law was not feeling well. While Shin spoke with her, Lucas turned to me and asked, "When we had that jellyfish yesterday, what did you think it tasted like?"

I cleared my head of all preconceptions, and just let a word come to the top of my thoughts. "Green pepper," I said, surprising myself with the answer. But as soon as it was out of my mouth, I knew it was correct. The outer side of the jellyfish was smooth, like the skin of a pepper. And the crunchiness everyone talked about was like what you get when you munch on a pepper slice. Add to it the nearly absent flavor and it was an animal doppelgänger for green pepper.

Lucas said, "That's exactly what I was thinking."

The ride to the airport was my last chance to ask Shin about jellyfish. I wanted to end where I started, with the question that first pulled me into the world of jellyfish: "Do you think the jellyfish will be winners in an acidified ocean?"

Shin said, "It's possible that they will do better, but certainly not settled. There's no solid data supporting it yet. I think the best answer is, we don't know."

Lucas pushed the question a little further: "A lot of animals are going to suffer as the ocean acidifies. Is it possible that jellyfish won't suffer as much?"

"Jellyfish are quite resilient, so it's plausible, but not proven," Shin replied. "It's interesting to look at the Bohai Sea, which is perhaps a template for the worst case. The Bohai is the innermost part of the Yellow Sea in China. Like the Ariake, it used to have very high productivity, but it has seen a ninety percent decline in harvest. Now only jellyfish are available." When I looked into the Bohai Sea later, I discovered that it is one of the most messed-up marine environments in China, a country that's not known for successful marine management. Disgraceful pollution has led to chronic blooms of harmful algae. Shellfish contain high levels of toxins. Coastal development and damming of the forty major rivers that flow into the sea prevent the migration of animals to their spawning areas. The dams also are increasing the salinity of the sea. Fish farms have introduced exotic species. Traditional fisheries, including herring and sea bream, are locally extinct. The list of problems goes on and on.

Shin continued: "I'm very interested in the fate of the Ariake Sea. Will the ecosystem remain jelly-dominated, as has happened in the Bohai Sea? What will happen to it next year, and the year after?" The question hung in the air. The Ariake Sea wasn't as messed-up as the Bohai, but would it be? When? There was no answer, not yet.

We were close to the airport, and I had one more question. "I know this is out of context, but why do people wear surgical masks here? It's not something we do in the U.S."

Shin looked surprised that I asked. "I guess for a couple reasons. It's protection against spreading germs when you or someone else isn't feeling well, but it also feels nice. If you aren't feeling well, it keeps your mouth and nose warm. It feels comforting."

We arrived at Fukuoka Airport, and I thanked Shin profusely—for putting together our itinerary; for his expertise, patience, and philosophy; and for his ability to turn any space into an educational opportunity. Lucas and I hugged and promised to stay in touch.

Inside, I checked the overhead monitors and found that my flight was delayed. After a long day that started at Yanagawa fish market, I was stinky from the spray of seawater and the goo of iridescent blue jellyfish. Since I had time, I stopped in the restroom to pull off my slimy tennis shoes and my dirty jacket. I bundled them in a plastic bag, hoping to shield everything in my luggage and everyone else on the plane from the smell. I switched to a sweater and clogs and roughly ran my fingers through my wind-bedraggled hair before approaching the impeccably dressed Japan Airlines check-in attendant. Nothing makes you feel grimier than speaking to someone who looks like a fashion model.

I bungled my way through security. My luggage needed to be inspected by this person, not that person. In contrast to U.S. rules, I was not supposed to check liquids. The sake from Tsushima that I'd bought for Keith had to be hand carried. Eventually I was cleared, and I found my way to my gate.

A shower would feel amazing, but it was still a flight, a bus ride, and a shuttle trip away. I knew I was on my way back to the real world, but I wasn't there yet. I cupped my hands around my face like a surgical mask. It did feel comforting to close my world in around my face with my still-briny palms. I took a deep breath, inhaling the sweet ocean scent of the Ariake Sea one last time.

About four hours after takeoff from Tokyo, I heard the announcement through my headphones. "Please make sure your seat belts are securely fastened. The pilot has informed us that we will be encountering significant turbulence. The captain has informed us that we will be passing through a strong weather front. The plane is equipped to handle this turbulence, but we ask that you remain in your seats with your seat belts fastened."

My seat belt was fastened, but I cinched it tighter. I felt the same rising anxiety that took hold of me on the ferry from Tsushima and on the flight into Rome on my way to meet Stefano and Nando. I clicked to the flight-path channel on the screen in the seat back in front of me. My stomach clenched. We were right in the middle of the Bering Sea, as far as possible from the safety of land.

I nervously clicked through the movie selections, finding the English version of *Rock of Ages*, and turned up the volume on my headset. "Heaven isn't too far away" did little to allay my fears, although Tom Cruise, sweaty and strutting in leather pants, was distracting. The plane jolted and rocked, but with the help of a couple of glasses of wine and with Foreigner blasting in my ears, I eventually managed to doze.

My return home felt wonderful. I heard updates. "We ate a lot of pizza." "The first day, we were tardy before Dad even got to the stoplight." "What did you bring us?" "My Greek gods test was hard." "I almost got my back handspring." "What did you bring us?"

Keith told me, "There's the list of all the things that needed to happen that you left me, and then there's the list you didn't leave me. The list that exists in your head and says things like under no circumstances can Isy wear the red skirt with the white shirt, or there will be a meltdown before breakfast." Now he knew.

The Halloween carnival at the kids' school was on Friday. They put on their costumes and played "Gangnam Style." A ghoul and a zombie did the horse dance around the living room. As we left the house, I felt the temperature dropping. Rain was in the forecast, so I forced the creatures to put on long layers underneath their costumes.

The weather change was the result of a strong high-pressure front pushing through the center of the country, driven by an unusually large trough in the jet stream. Reports started to trickle in about a megastorm building on the East Coast. A tropical depression named Sandy was driving north. Usually, a storm on this track would head out into the ocean, where it would dissipate, perhaps without even making landfall. But two factors stood in the way of a slow burnout at sea. An unusually large trough in the jet stream carried winds that mixed with Sandy, intensifying the storm. And another front parked off of Greenland was stubbornly holding its position. The official weather term for what this Greenland front was doing is blocking. Just as a defensive football player blocks the movement of the opposing team, this front blocked the offshore movement of Sandy, now intensified by the cold front.

The timing of the storm so close to Halloween spawned the name Frankenstorm, which implied the role humans may have had in creating this monster. For the first time I could remember, the effect of climate change was conspicuously cited as a reason for the power behind the storm. Later analysis showed that that attribution was debatable. But climate change probably did have a role in building the blocking system that pushed the storm's track back toward the East Coast, where it caused enormous damage. Scientists are still studying whether such blocking systems will become more regular and spawn more Frankenstorms. However, it is clear is that warming seas will make hurricane seasons longer and furnish more energy to push storms northward out of the tropics. In addition, rising sea levels will mean that, when hurricanes do happen, there will be more water that can be pushed farther onshore, causing more damage.

A few weeks later, I was at a family fun center in a strip mall in north Austin. It sucked. And I know it wasn't just me; my kids hated it too. They hated it the way someone on a low-carb diet hates the smell of fresh bread: It's so enticing and yet so bad for you. Isy was stuck deciding whether to spend her prize tickets on the cheap, sparkly headband she wouldn't even remember owning once she got to the car or on the sour candy that made her mouth pucker. The decision was torture. So the place sucked. And Isy was on her sixth circuit around the confines of the store, evaluating choices, becoming more and more frustrated, adding, subtracting, agonizing, eventually breaking out in tears. By this time, I was so overstimulated from the *ping*s and *zing*s and *zap*s of the arcade and so disoriented by the false darkness in the middle of the day that I was circling too, avoiding the orbit of my daughter until the excruciating decision was complete.

On about my fifth lap, I noticed one item that didn't glow or sparkle. It wasn't fluorescent. It wasn't even wrapped in plastic. Stuck on a corner shelf was a globe about the size of a grapefruit; instead of being held on its axis by a rod stuck through the poles, the sphere was suspended on its stand by a magnet at the North Pole. A band of copper traced the equator. This tiny earth spun easily on the magnet at the top, and I gave it whirl. After I returned from Japan, I'd found an archive of global weather where I could watch the swirling systems that dance around our globe each day. With my finger I now retraced the path of the powerful front that rocked my ferry ride from Tsushima. I continued to follow it through the stormy Ariake Sea and then across the Pacific Ocean, where it intersected with my airplane. At the coast of North America, the front withered, but before it turned northward, it handed off its energy like a baton in a relay race to a newly formed continental front. This front worked its way over the Rocky Mountains, gathering energy in the flatlands to the east. It strengthened as it passed over me and my kids at the school Halloween carnival in Austin. A few days later, it slammed into Tropical Storm Sandy and stalled, unleashing horrifying damage to both life and property. Amid the artificial chaos of an arcade, this tiny spinning globe was transformed into an unbidden tangible reminder that we are all along for the same ride on this small planet we share.

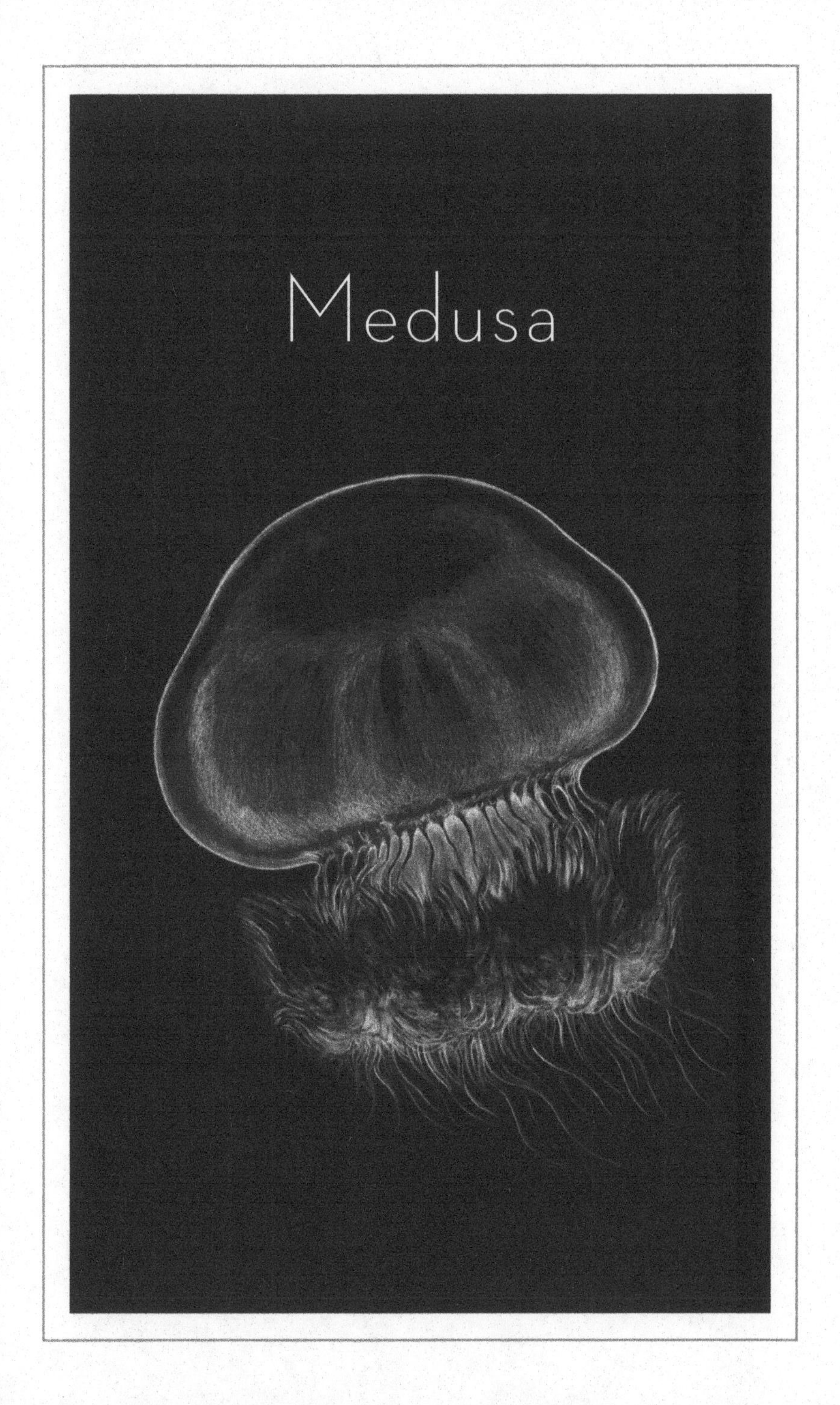
Medusa

Toxic Cocktail

"We're thinking about having the bar mitzvah in Jerusalem," my sister told me, talking about her son's coming-of-age celebration. *No, no, no, please don't,* was my first thought. I did not want to go back to Israel. True, I had discovered my love of the ocean there, but I'd also discovered unbearable conflict and anger, not to mention how fat I could get drowning my depression in pita filled with hummus and falafel. Even after three decades, I didn't want to return.

But it wasn't just my family that was pushing me back to the first place where I'd peered under the sea. The jellyfish were doing their part too. Over the years, my Google alerts had filled with stories of jellies in Israel, which has become a hot spot for jellyfish-related problems as well as innovations. I'd read of scientists creating superabsorbent paper towels from jellyfish, jellyfish bandages, stinging-cell drug-delivery devices and cosmetics, and a jellyfish sting–blocking sunscreen. One jellyfish in particular, the nomadic jellyfish, had hit my Google feed heavily with stories of massive blooms clogging intake pipes at power plants along Israel's Mediterranean coast.

The problems started in the 1980s, when swarms of round white jellyfish began appearing. They reminded Israeli scientist Bella Galil of a similar jellyfish first noticed off Israel's western coast in 1977. Bella identified the animal as a new species, probably native to the Indian Ocean, and surmised that it had traveled to its new Mediterranean home through the Suez Canal, a 120-mile channel from the Indian Ocean to the Mediterranean Sea. Completed in 1869, the canal was one of the greatest technological achievements of the era, saving sailors months of time in rough seas and shipowners piles of money by quicker transit of their goods. Instead of traveling all the way around Africa to reach Europe from Asia, ships could take a shortcut through the Red Sea.

To capitalize on the existing geography, engineers designed the canal to pass through two large inland lakes, called the Bitter Lakes because of their high salt content. It's thought that the extra-salty water also acted as a biological barrier, preventing creatures from migrating between the newly connected oceans. The nearby Nile also acted as a deterrent, flooding the north end of the canal with fresh water and hindering the migration of marine creatures that couldn't handle low salinity.

However, over the century and a half since the canal was built, the salt was washed out of the Bitter Lakes. Then, fifty years ago, the completion of the Aswan Dam plugged the freshwater flood of the Nile. The Suez is now uniformly saline and uniformly hospitable to marine creatures. And they have been using it as a superhighway. The number of new invasive species in the Mediterranean has doubled in the last two decades. Over four hundred species have traveled the Suez from the Red Sea to the Mediterranean and established populations there. Together they have earned their own name, Lessepsian migrants, which recalls the Frenchman who envisioned and developed the Suez Canal, Ferdinand de Lesseps. Because of the flow pattern in the canal, only a few species have made the reverse passage from the Mediterranean to the Red Sea.

Some of the newly arrived species benefit humans. Commercial fishermen in the eastern Mediterranean routinely haul up nets containing narrow-barred Spanish mackerel and tiger prawns, both originally from the Red Sea. Other species, finding few predators in their new habitats, have had negative effects. Two rabbitfish—so called for their small rabbitlike mouths, though their herbivorous appetites and quick reproduction cycles make the name even more apt—have razed hundreds of acres of rich seaweed beds along the coast of Turkey. The destruction has had a domino effect on the ecosystem. As the scientist who studied the seafloor clear-cutting by the rabbitfish told me, "When you remove the forest, all the birds are gone." Another problematic migrant is a poisonous pufferfish originally from the Red Sea. Related to the Japanese fugu, which must be prepared by a master chef to avoid poisoning the diner, this pufferfish is now one of the ten most abundant fish caught in the eastern Mediterranean. Locals unfamiliar with the animal have been hospitalized after eating it. And the blue-spotted cornet fish, a predator that uses marine protected areas as refueling stops, is racing its way across the Mediterranean. It may be the first Red Sea species to reach Gibraltar and the Atlantic beyond. Fish scientists call it "the sprinter."

In the Mediterranean, blooms of the nomadic jellyfish are now massive. During one such huge bloom in 1990, surveys found a "jellyfish belt" a mile offshore with as many as twenty-five nomadic jellies per cubic meter; that's like twenty-five water balloons packed in your oven. All those jellyfish stalled operations and increased costs at power plants. Fishermen had to stop fishing because their nets were damaged by the weight of the jellies and their catches were ruined by the jellyfish's toxin. Beachgoers stopped swimming because the stings caused a nasty burn in some people and high temperatures and muscle aches in others.

When I was in Italy, Stefano Piraino and Nando Boero had shown me an article they had just coauthored with twenty-two other scientists called "Double Trouble." The article explained that the Suez Canal was about to get a huge renovation, one that would double its capacity. The doubling worried them. Not only would it double the number of ships that could travel the canal, but it would also greatly increase the amount of water flowing through it and the number of creatures that could swim through it. It was impossible to say what impact the new arrivals would have on Mediterranean ecosystems.

Typically such a huge construction project in a marine environment requires an assessment of the risks and a plan to minimize problems if they arise. The "Double Trouble" scientists noted that three signed treaties required just such an assessment, but none of those treaties had been enforced, and no assessment was being performed. Ideas for mitigation existed: installing barrier-like locks or reestablishing the original salinity barrier of the Bitter Lakes could slow or halt the flow of species between oceans.

Reading the article, I was surprised by the story. I hadn't heard anything about it in newspapers or online. That was exactly the problem, they said. No one was paying attention. They urged me to call Bella Galil, whose name I already knew from her work on the nomadic jellyfish. She was the lead author on the "Double Trouble" article. When I returned home, I did call Bella in Haifa. She told me that $8.5 billion had already been raised for the project and ground had been broken on the canal expansion.

"How can that be?" I asked. "Why haven't I heard anything about it? And why aren't the responsible agencies enforcing their treaties?"

"Why don't you ask them yourself?" she shot back.

So I tried. I spent two weeks e-mailing and calling the United Nations agencies whose treaties had been signed and whose job it is to protect the oceans and their ecosystems. The people at the Division for Ocean Affairs and the Law of the Sea never picked up the phone. They responded to my e-mails with an automatic message. Someone at the Barcelona Convention picked up, said he had no response, and promptly hung up on me. After I was bounced around among the administrative assistants of half a dozen people around the world, someone at the Convention on Biological Diversity said he was sorry he couldn't talk—he was busy sitting on a tarmac somewhere—but he e-mailed me a statement indicating that he and his cohorts would be taking no action on the expansion. All of this was in violation of the treaties they oversaw.

It was a moment for me to straighten my spine. I did something I never could have imagined doing when jellyfish first swam into my landlocked world. I pitched the story to the largest media platform I knew, *The New York Times*. I was overjoyed when they accepted the piece, but I'd had a feeling they would want it. This story needed to be told. When the piece was published, the accompanying artwork showed jellyfish swimming under the ships in the Suez Canal.

Bella and I both thought that once the story hit an international stage, once the world heard about the lack of environmental oversight on such an immense project and about the broken treaties in one of the biggest hot spots for invasive animals anywhere, something would change. I thought the responsible agencies would step up and do their job. Or perhaps a non-governmental agency or a watchdog group would step in and make noise about it. But I was naive. Despite other major news outlets picking up the story, the international community remained mute. And despite Bella's continued efforts—she collected several hundred signatures from scientists around the world protesting the lack of oversight and also petitioned the European Parliament—no environmental assessment was performed. No plans to prevent the spread of invasive species were developed.

The Suez Canal expansion was completed in record time. The project was supposed to take five years, but just one year later, military planes flew overhead during the ceremonies in which Egyptian president Sisi inaugurated the New Suez Canal. A massive amount of water with all the species it carried with it was already flowing freely through the enlarged channel. Bella said, "We are playing a Russian roulette, not with a bay or a river but with the Mediterranean Sea." I had never met Bella in person, but that was another reason to return to Israel.

While the biggest jellyfish problem in Israel was certainly the nomadic jellyfish, beachgoers were also encountering stingers like the sea nettle, the mauve stinger, the man o' war, and a box jellyfish. Of all the jellyfish in the sea, the box jellies are the most toxic. The Pacific species,

Chironex fleckeri, may even be the most toxic animal in the world. It is said to be able to kill a person in under three minutes. Numbers of fatalities are hard to pin down, but are often estimated to be dozens every year, more than what is attributed to sharks. Another group of about twenty-five species of box jelly thought to exist in oceans around the world cause Irukandji syndrome, which involves severe pain, nausea, difficulty breathing, sweating, anxiety, and often "a feeling of impending doom." Despite the horrific symptoms, as of 2013 only a few confirmed deaths had been attributed to Irukandji, although many others are suspected. Unfortunately, warming ocean waters may be expanding the range of these jellies.

Hawaiian scientist Angel Yanagihara has been studying the potent virulent box jellyfish toxins since one predawn morning in 1997. A swimmer all her life, Angel was looking forward to a long workout in the warm tropical water near Waikiki. Angel passed an older woman. They smiled at each other and the woman noticed Angel's goggles and swimsuit.

"Don't go swimming," the woman cautioned, pointing at clear blobs in the sand. "These are box jellies. They are dangerous. Stay out of the water."

Angel walked on. A petite woman with her brown hair pulled back in a ponytail, she had just finished her doctorate in biochemistry. She figured the nylon swim shirt she was wearing would be protection enough. And if she were stung, well, she'd suffer a little pain, a little redness. Despite all her experience in the water, she wasn't aware of how toxic box jellies are. Angel blew off the old woman's warning.

She waded into the surf, adjusted her goggles, stretched out her arms, and kicked off. When Angel was a child, her father, a military officer, had encouraged her connection to the sea. He strapped a scuba tank on her back when she was six, before she was even big enough to walk with it. Today, as always, she felt powerful and calm in the ocean.

At the last leg of the hourlong swim, as she was fighting a current to swim back to shore, a burning knife cut into her neck. The pain was like nothing she had ever felt, intense, hot. Angel dived, changing depths to escape whatever was attacking her at the surface. She recalled the old woman's warning: the box jellies. Resurfacing, she inhaled, but she couldn't catch her breath. It felt like water was filling her lungs. She had surfaced in a second cloud of box jellies.

Sick and suffering, Angel dragged herself onto the beach. Her throat was closing; her breathing was labored. She pulled off her swim shirt, and doused herself in the cold, fresh water of a beach shower. This was the worst thing she could have done. Cold, fresh water hastens the eruption of box jelly stinging cells. She passed out. It's likely that she had Irukandji syndrome.

Angel woke in an ambulance. Tubes of lifesaving liquids and gases snaked into her body. A woman who lived in a nearby apartment had called the paramedics when Angel lost consciousness. Now the same woman was telling the responders about the box jellies, urging them to bathe Angel in hot water. This local knowledge was spot-on. As the responders soaked Angel's wounds in hot water, she began to feel some relief. Years later, Angel would show in the lab that the main attack proteins in box-jellyfish toxin are heat-sensitive. Twenty to forty minutes in hot water deactivates them permanently.

After this experience, Angel began reading up on the biochemistry of box-jelly stings. She didn't find much. The medical literature contained glaringly little about box-jellyfish toxins. She did learn that the genus that likely delivered the sting was *Alatina*. *Alatina*'s bell is a clear rectangle about five inches long, with white bands extending down the four long corners, ending in four pale pink tentacles that contract and extend from just a few inches to three feet when the animal fishes. *Alatina* is Latin for "little winged one," like an angel. The coincidence of the jelly's name and her own wasn't lost on Angel, who has since worked to understand the animal's secrets.

Alatina probably lives most of its life offshore in the deep water, but at a cue from the light of a full moon, mature adults swim upward and landward. About ten days later, they congregate on south-facing beaches during the early-morning ebb tide—the exact time and place Angel had picked for that fateful swim. Once fertilized, *Alatina* broods her developing eggs on a strand inside her bell, like a string of pearls. The mating ritual has since become useful to Angel, who can depend on finding box jellies for her research. Eight to ten days after a full moon, she and the members of her lab walk the shallow waters of Waikiki Beach during the predawn ebb tide. They collect the box jellies that swim toward the flashlight beams they shine in the water.

Angel extracted a cocktail of about four hundred different molecules from *Alatina*'s stinging cells. Because jellies are a very old group of animals, she reasoned that their toxins might be very old too. One family of ancient toxins are called porins, which are found in bacteria, mold, and fungi. Porins, as their name suggests, form pores in the outside membranes of cells. Like molecular buckshot, an attack of porins causes cells to literally spill their guts. For organisms like bacteria, fungi, and molds that absorb—rather than eat—nutrients from the world around them, this attack strategy surrounds them in food.

In 2001, Angel identified a porin from the box jellyfish that had stung her. The porin molecule has a corkscrew-like structure that can bore into a cell membrane. Six to eight porin proteins form a ring, like a small beaded bracelet. Electrical charges on either side attract the proteins to one another so that the porin can self-assemble, as if the beads had magnets on either side that snapped together. Once assembled, the porin ring then comes to rest on the outside of a red blood cell, and like a round cookie cutter, drills into the cell membrane. The space in the porin's center becomes the hole through which the insides of the cell pour out. Flooding an animal's body with porins results in a quick death. It's actually the potassium that leaks out of blood cells that causes the damage. So much of this ion enters the blood that it can cause cardiovascular collapse. "The porin could cause death in seconds in a mouse, far faster than any snake toxin ever discovered," Angel said when I spoke to her by phone. This ability to form holes in blood cells is reduced when it is heated to just 115 degrees Fahrenheit.

The temperature sensitivity explains why dousing with hot water, not cold, stops the pain of a box-jelly sting.

Researching old cures, work that reached back to the 1880s, Angel discovered that certain metallic salts sometimes decrease the activity of toxins. Angel and her colleagues tested the therapeutic effects of some of these and found that zinc gluconate, which is used as a lozenge for the common cold, was the most promising. In test tubes and in mice, it stopped jellyfish porins from riddling red blood cells with holes. The reason, Angel believes, is that zinc gluconate is highly attracted to those very spots on the porin subunits that stick together to form a ring. It attaches to the subunit faster than the subunits can stick to each other, so they never circle up. If they can't form a ring, they can't make a hole in the red blood cell.

Soon after Angel published her work on jellyfish porins, she received a surprising e-mail from champion endurance swimmer Diana Nyad. In her twenties, Diana had made headlines for long-distance swimming when she swam around Manhattan in the span of a workday. A few years later, she swam from the Bahamas to Florida, a distance of 102 miles, in twenty-seven and a half hours, setting a world record for endurance swimming for both men and women. Now in her sixties, Diana had made it her goal to be the first person to swim from Cuba to Florida without a shark cage.

Diana had tried to make the historic swim three times already but had failed. She hadn't faltered for physiological, athletic, or even psychological reasons. She trained for years and spent innumerable days and weeks and months in the near–sensory deprivation of a long-distance swim. She could keep herself awake for the mind-numbing fifty or sixty or seventy hours the swim would take. Diana could manage the brutality of seasickness and of throwing up over and over from swallowing so much seawater. She was able to overcome the hunger and the cold and the grueling pain of the seawater that chafes not just the skin, but the mouth and throat too. She hadn't even failed because of sharks. Diana had failed because of jellyfish. She'd been so brutally stung by box jellyfish on her third attempt that she almost died. Her arms, legs, back, and neck were lashed by the fiery tentacles of box jellies, which caused her to feel as if her entire body was dipped in hot oil. Over the phone, Diana told me about the incident: "If you are going to go out in the open ocean, all kinds of things are going to be out there, known and unknown, and you are just going to have to accept that as part of the sport and the adventure that you're a part of. But I must say, when I was stung by those box jellyfish. . . . I wouldn't ever wish that on my worst enemy. It was the stuff of science fiction."

Steve Munatones, who was the official observer for Diana's swim and is the founder of the World Open Water Swimming Association, explained that today the sport of open-water swimming is often about jellyfish. He said that while it's true that swimmers have pushed the envelope, going to places farther offshore, he didn't think that the incursion of people into the realm of jellyfish told the whole story. "There have been many others besides Diana from the 1950s on who have tried to swim from Cuba to Florida. No one that I know of up until probably the 1990s was running into jellyfish, and certainly not box jellyfish."

After that attack, Diana realized that jellyfish were an unavoidable part of her swim from Cuba to Florida, so she contacted Angel to come up with countermeasures. Angel developed a predictive model of the times when the moon, the sun, the tides, and the depth of the seafloor aligned for a higher likelihood of box jellies swarming. Like much of the world's plankton, box jellies make a twice-daily vertical migration, more so at high tides than low tides and even more eight to twelve days following the full moon. Angel designated these periods of greatest box-jelly risk "no-go" times. Working with a technical swimsuit company, Diana had been developing jellyfish armor to minimize discomfort and drag. Angel tested and retested the design of the weave so that jellyfish stinging cells wouldn't pierce the material. Diana had a mask constructed that covered her entire face, including her lips; it even reached inside her mouth like a retainer. The mask was horribly uncomfortable, changed her breathing, and forced her to swallow even more seawater than open-ocean swimming normally did. It was like swimming in a raincoat and made her throw up more frequently.

Like so many of us who swim in the oceans where jellyfish are, Diana also wanted a treatment in case she was stung. Given the frequency of contact between beachgoers and jellyfish—an estimated 150 million stings a year—it is surprising that there is no medical consensus on how to treat jellyfish stings and that protocols vary widely. Some say to use hot water, others cold water. Some vinegar, others alcohol, and still others baking soda. And peeing on stings continues to have its place in the rumor mill.

The reasons for the shortfall come down to three big problems. First, there are many different types of jellyfish, and each may have different sets of toxins, which may require different treatments. Second, you might not always know which jellyfish stung you. Last, there really hasn't been a comprehensive, rigorous study on the problem, although that might be finally changing. In 2016, jellyfish scientists, including Angel and others from Israel, Australia, Europe, South Korea, and the United States, set up a collaborative network to develop a universal sting protocol. Some results are already emerging. Hot water disables the toxins that cause pain in all jellyfish so far tested. Vinegar appears to be an effective treatment too. If you are stung, your best bet is to pluck off any visible tentacles using tweezers if you have them. Then douse your sting in vinegar and then hot water. Cold water will make the sting worse. Alcohol, baking soda, and urine (except for the fact that it's warm) are ineffective.

Several different groups are also working on antidotes, though most are in early stages. The Chinese have developed a treatment cream called JSM, which they have been testing at the beach in Qingdao where giant jellyfish can be a menace. However, the treatment still needs to undergo rigorous clinical trials with good controls. A French pharmaceutical company is also developing an antidote to two Mediterranean jellyfish with bad stinging cells.

When Diana called, Angel had ideas for an antidote based on her research with zinc gluconate. She formulated a chartreuse paste that contains zinc gluconate and proprietary components that are under patent review. Before Angel turned the formulation over to Diana, she tested it—in an act of supreme confidence that terrifies me—by laying an entire box jellyfish across her arm as medical personnel stood by. Once the pain started, she removed the jellyfish and smeared her arm with the formulation. The sting abated. The searing pain vanished. The formulation worked.

In 2013, a window opened that looked favorable for Diana's swim from Cuba to Florida. Clear weather was forecast, and the water was still warm enough to make the swim possible. Oceanographers modeling currents predicted that the position of the Gulf Stream would help push Diana toward Florida. Angel's jellyfish models showed that the time wasn't optimal, but it wasn't no-go, either.

Early on the morning of August 31, 2013, Diana left Cuba accompanied by a thirty-five-member support team that included Angel. At nightfall before the jellies rose from the deep, Diana paused in the water. Her team handed her the full-face mask, and she put it on. When she was a child, Diana's father told her that the name Nyad comes from the word *naiad*, a swimming nymph of ancient Greek mythology. The mask's thick translucent skin, which distorted her human features, seemed to turn Diana into a sea creature herself. Throughout the night, Angel patrolled the waters around Diana, repeatedly free-diving down fifty feet or more, scoping the water for box jellies that might be rising to the surface. The first night, there were none.

The second day, the seas were choppy, but everything else was working in Diana's favor. The currents were pushing her toward Key West in record time. There were no storms, no sharks, no jellies. The second night, Angel was in the water with Diana again, diving below her and around her, scanning for stingers. By very early morning, Angel climbed onto the support boat to take a nap. Soon after, Diana's swim coach told her she was on pace to reach Florida by early afternoon. Feeling safe, Diana pulled off her protective mask and her full-body suit. Through her sleep, Angel heard a voice on the radio saying Diana was swimming in just her swimsuit. *No!* Angel thought. It was sunrise. It was the tenth day after the full moon. The tide was rising. This was a no-go situation.

Angel bolted out of bed and grabbed her mask and snorkel. She told the other safety divers, "Get in the water! Look out for box jellies!" She jumped in the water herself, and just as she and Diana came over the top of the reef on their way into Key West, Angel saw the box jellies. She reached into the water with her bare hand, grabbed one by the bell, and snatched it out of Diana's path. "We managed to get her ashore safely," Angel said.

When Diana Nyad reached the Florida Keys just under fifty-three hours after she left Cuba, she could barely stumble up the beach. Onshore, she told interviewers that with each stroke, she imagined pulling Florida toward her and pushing Cuba away. The mantra that kept her going during the epic swim was "Find a way. Find a way." With those thoughts she accomplished what no other human being has ever done, an idea transformed into an inspiring act of skill and will. But also, the sea nymph had the protection of one unflappable guardian angel with a deep understanding of the fearsome box jellyfish.

While the toxins that make up the box jellyfish are particularly dangerous, they are really only the tip of the iceberg when it comes to jellyfish poisons. In Israel, I would have the chance to meet Tamar Lotan, a scientist from the University of Haifa. She and her colleagues had looked at the ingredients in the toxic cocktails of three different cnidarians: moon jellyfish, a sea anemone, and a freshwater polyp called hydra. They discovered not tens or scores but hundreds of different proteins in each stinging cell: 743 in the moon jellyfish, 380 in the sea anemone, and 297 in the hydra. How many of those hundreds of proteins were common to all three animals? Only six. Nearly 1,400 different proteins were found in the stinging cells of just three cnidarians. And not just the number but also the destructive functions of the molecules are numerous. Some break apart proteins; others break apart the molecules that make up cell membranes. Some block the signals between neurons. Some attack muscle cells. Many break apart red blood cells. Some have similarities to toxins in brown spiders and snakes. Some are unlike any other toxin we know. A recent review on the composition of cnidarian toxins drew the understated conclusion that "there are large gaps in our knowledge about the toxins and their specifics." One thing is certain, however. There's an enormous amount of unexplored biochemical richness inside the stinging cells of cnidarians. It's a research area ripe for both discovery and solutions.

Sting Block

There were the crying babies in the row behind us, but they were to be expected. An egg-shaped flight attendant hustled by, asking in an obnoxious way, "Kosher meal? Kosher meal? You, kosher meal or not?" A disheveled man was walking the aisles trying to put together a minyan, a quorum of ten men, for a prayer service in the morning. In the midst of it all, a woman with wild curly hair wearing a too-colorful poncho fixed laser-beam eyes on me and asked, "Are you a doctor?"

Who, me? I looked to either side as if I weren't sitting with my family. She meant me. I said, "Is something wrong?"

"I just need someone to give me a shot." She waved a black case containing needles and vials at me. "I thought you might?" She thrust them toward me.

"Uh, no. Sorry," I answered. "I'm not a doctor." Not that kind anyway. "I'm no good with vertebrates," I muttered to Keith. She turned away, looking down the rows for a human doctor.

"It's like a *Seinfeld* skit on this plane," Keith whispered back.

I leaned back against the seat. "But free wine," I said, raising my plastic cup. I took a sip, calming not just my flying jitters, but my apprehension about returning to Israel. My feelings about this trip were not at all crystallized. I was looking forward to the ceremonies in Jerusalem with my family, and I was excited to meet the jellyfish experts I'd contacted. Also, my Google alerts had reported a large bloom of the nomadic jellyfish, so big that it threatened to shut down a power plant south of Tel Aviv. If it stuck around, I might be able to see it. I hoped this trip would somehow be a "restorative process," as my next-door neighbor, who is a therapist, suggested to me. So far, though, Keith's analogy was dead on. I felt I was headed toward frustration with occasional moments of awkward comedy.

When we arrived in Jerusalem, my nephew's bar mitzvah was still two days away, so of course I went to meet with jellyfish scientists while Keith and the kids went on an archaeological dig in caves that were inhabited two thousand years ago. Tamar Lotan, who had looked at the diversity of cnidarian toxins, and her husband, Amit Lotan, a molecular biologist who works on understanding just what makes stinging cells sting, live in Tiberias, about a two-hour drive north of Jerusalem. Amit had founded a company called Safe Sea, which produces a sunscreen he claims blocks jellyfish stinging cells from deploying, and I was headed to the company headquarters.

The rental-car people chose a hot magenta number for me, and I felt more than a bit conspicuous as I made my way doubtfully around Jerusalem's windy streets, following the instructions of an uncertain GPS that came with the car. I was brought to tears only once, when I missed a turn and ended up in a dirt parking lot on a cliff in back of a grocery store. Eventually the navigation system and I came to an understanding, and I found my way onto Highway 1 heading north, which skirts the West Bank on its way to Tiberias.

The bright new offices of Safe Sea were all glass doors and sleek lines, with a fabulous view of the sparkling Sea of Galilee, along whose shores Jesus is said to have spent years teaching his disciples. Amit Lotan looked vaguely like a mid-career James Taylor, tall and thin with an angular jaw, bright eyes, and a bald spot encircled by a curtain of hair that covered his ears. After setting a container of the plumpest dates I'd ever seen on the table between us, Amit told me that the idea for his sting block originally came from the clownfish, that cute orange-and-white fish that Disney named Nemo. Thinking about how clownfish snuggle down into the arms of anemones and even lay their eggs there, Amit decided that "if the clownfish knows how to stop the sting, I could find out how too."

I've come to believe that the stinging cell is the most sophisticated cell to have ever evolved in any animal. I've heard it said that the vertebrate eye is the most complicated organ, and critics of the idea of evolution point to the eye as something too sophisticated to result from the haphazard process of natural selection. And it's true that the eye is an incredible organ. But it's imperfect. The optic nerve sits right where light hits the retina, making a blind spot. Once I began to learn what a complex and highly sophisticated biological construct the cnidarian stinging cell is, I was completely floored. If you want to wax mystical, in my book the stinging cell's a much better option.

An unexploded stinging cell is minute: between 2 and 250 microns long, in the range of the width of a human hair. If you look through a microscope at an unexploded stinging cell, you see a round or oblong capsule mostly filled with liquid; this capsule has a hinged trapdoor at one end.

The skin of the capsule is a remarkable material made up mainly of two proteins. One is called minicollagen, which is like the collagen that gives our skin its elasticity. The other is a protein that has only recently been identified, called cnidoin because it is found only in cnidarians. Cnidoin is even stretchier and stronger than collagen. When tugged, cnidoin can grow 70 percent in length without breaking. At the molecular level, it bears a strong resemblance to spider silk. When combined, minicollagen and cnidoin spontaneously form sturdy thin sheets that have unparalleled strength and spring. The sheets are also porous, allowing water, but not larger molecules, to pass through them.

Inside the capsule is a very long, hollow microscopic tubule adorned with barbs, sharp arrows, or a combination of such stabbing devices as well as toxins. The entire tube is inverted like a leg of a pair of tights turned inside out, and then twisted up tightly. "It's like a twisted-up rubber band," Amit explained. Just how the cell creates so much twisting when it builds a stinging cell is a mystery.

Besides the tubule, the inside of the capsule is filled with water molecules, long chains of a polymer called gamma-glutamic acid, and a lot of calcium ions. The calcium has the special molecular quality of being able to hold on to several of those polymer molecules at the same time, like holding hands with other people. When the stinging cell is triggered, electrical changes in the cell signal the calcium ions to release their grip on the polymers, like dropping hands. In that instant, the number of molecules inside the capsule increases many times, making the concentration of the water inside the capsule much lower than outside. Water rushes inside to equalize the concentration, expanding the super-stretchy capsule skin by 30 percent. The swelling jacks up the capsule with an unfathomable amount of pressure: 150 atmospheres. Right now, the air pressure pushing down on you is 1 atmosphere. You inflate your car tires to just over 2 atmospheres. A stinging cell holds seventy-five times more pressure than a car tire.

When the trapdoor at the end of the stinging cell is triggered open, all that jacked-up pressure and the elasticity of the capsule skin forces the tubule out. The tubule, which was twisted inside out, unfurls like the leg of a pair of tights being turned right-side out, elongating as much as one hundred times the length of the capsule. The tubule's sharp spears lodge into its prey and toxins pump from tiny holes at the end of the hollow tube. It wasn't until we developed a high-speed camera that could film over 1.4 million frames a second that we could capture just how fast the stinging cell deploys. It takes just 700 nanoseconds to fire. The pointed dart hits its prey with a pressure of over a million pounds per square inch, similar to a bullet fired from a gun. A Ferrari can accelerate from 0 to 60 mph in three seconds, an acceleration of 3 *g*, where *g* means *g*-force, or the acceleration due to gravity. The acceleration of the stinging cell is 5 million *g*. It's thought to be the fastest motion in the animal kingdom. I told you, the stinging cell is incredible!

While the stinging cell is unique to jellyfish, anemones, and corals, each of these groups has come up with an astonishing number of riffs on the theme—about thirty different types. The greatest differences are in the tubules themselves, many of which look like a microscopic exhibit on weaponry. The spines can be uniform in size or can vary from large and toothy to small and needle-like and back again along the length of the tube. In some sections they are closely set, almost furry, like the coating on a burr that snags you as you walk through the woods. Elsewhere they can be wide-set, like sharp spines on cactus pads.

From Amit's point of view, it was one thing to understand the mechanics of stinging-cell explosion, but to make a product that blocks it from happening, he needed to understand the detonator. If he could defuse the bomb, he'd be in business. Amit told me about a series of ingenious experiments from the 1980s and 1990s by researcher Glen Watson and his colleagues, who learned what pulls a stinging cell's trigger. After I returned from Israel, I contacted Glen, who has been working on stinging-cell triggering for almost three decades. Most of our understanding of stinging-cell detonation comes from Glen's work on anemones. The details are probably a bit different in jellyfish, in the way the password to unlock an Android is different from the one for an iPhone, but the fundamental systems are thought to be similar.

It's important to remember that the primary purpose of a stinging cell is not to sting *us* but to capture small zooplankton. Referring to Charles Schulz's comic strip *Peanuts*, Glen said, "You can envision the prey as sort of like Pig-Pen. . . . Things are flying off the prey as it swims, and what is coming off the prey is mucus. The mucus contains proteins that are renowned for being attached to really long-branched sugar networks. Because it would be the first signal of prey that an anemone would detect, we used that as a stimulus." By adding increasing amounts of these sugars to the tank, chemically simulating what an anemone would "smell" if a Pig-Pen-like zooplankton prey came closer and closer, Glen and his colleague tested how sensitive the stinging cells were. As expected, when the sugar concentration increased, the number of stinging cells that fired also increased—up to a point. Then the anemone did something unexpected. As the sugar concentration continued to climb, the firing *de*creased. The decline indicated that there was more than just smell tripping the trigger.

Glen and his colleague suspected the second triggering signal was sound. They had long known that the stinging cell's trigger is a clump of stiff cilia called a hair bundle, which looks strikingly similar to the hair bundles found in the cochlea inside our ears. When a fly flies, its wings vibrate molecules in the air. The air molecules bump up against other air molecules and transfer those vibrations away from the vibrating creature toward our ears. When the wiggling air molecules hit the hair bundles in our ear, the hair bundles wiggle back and forth, sending signals to our brain, which we interpret as annoying buzzing. A mosquito beats its wings faster than a fly, so the hairs in our ears wiggle faster and we hear that vibration as a whine—higher frequency, still annoying. When zooplankton swim, they wiggle their legs and antenna, making insectlike vibrations in the water. To test if stinging cells were using their hair bundles to hear these zooplankton, Glen developed an instrument that vibrated a small piece of gelatin-coated fishing line at different frequencies. He tuned the frequency like tuning a radio station to create different zooplankton buzzes. He could then count the number of stinging cells embedded in the sticky fishing line.

By combining the frequency tuning with different chemical signals, Glen unraveled the stinging cell's trigger. He discovered that the stinging cell has layers of controls, like having multiple passwords on your phone. If the animal hears a buzz but doesn't smell the sugars that zooplankton emit, no initial attack occurs. If an animal smells zooplankton but no buzz follows, there's no detonation either. Only if both smell and buzz occur will the animal fire an initial round. And then, the stinging cells get even more finely attuned to the battle. If the animal detects the smell of zooplankton blood in the water and hears a higher panicked sound frequency of an injured prey, it will fire a second volley of stinging cells. But if the stinging cell doesn't smell blood after a first attack, it's a signal the first volley wasn't successful, and the animal holds its fire.

Such astonishing sophistication in the wee trigger of a stinging cell seems almost unimaginable until you think about what it's guarding. As I hope you'll agree at this point, the stinging cell is an unimaginably complex weapon, what with its crazy barbed dagger, cache of toxins, superstretchy strong capsule, and high-pressure-based deployment scheme. The drawback is that it's a single-use piece of equipment. Stinging cells can't be reused. Once fired, the animal has to invest in building a whole new piece of elaborate artillery. When you think about the value of the asset, it makes sense to arm it with a high-tech trigger.

Some of the most intriguing work on stinging cells is just now happening, and it may help repair damaged hearing. Loud noises create large sound waves that physically rip apart the hair bundles in our cochleas. Hair bundles don't work in disarray, and we mammals have a very limited capacity for hair-bundle repair. Anemones, on the other hand, produce a mucus that contains proteins that actually rebuild their hair bundles. We don't make these proteins, but we might be able to steal them from anemones. Glen and his colleagues treated damaged mice cochlea with anemone repair proteins, and their hair bundles healed in just an hour. Glen told me that a potential treatment for human hearing loss would take another decade of testing, but the research reminds us that we can learn unexpected things by tuning in a little more closely to the secrets of the sea.

Back in the Safe Sea offices in Tiberias, Amit explained how the ingredients in his sting block stymied four of the mechanisms that deploy the stinging cell. First, if the tentacle doesn't touch your skin, it can't fire stinging cells into you. So he added silicon, creating a slippery surface between a swimmer and the water, decreasing the possibility of a tentacle contacting the skin and increasing the probability that it slips off if it does. Next, like brine shrimp, we humans have Pig-Pen-like tendencies in the water. We emit a dust storm of sugars and proteins that signal our presence. So Amit incorporated chemicals that can chemically bind to the same sensors on the stinging-cell trigger that smell the sugary secretions of the prey. These blocking chemicals essentially give the stinging cell a stuffy nose. Along the same lines, scientists have long noticed that when jellyfish bump into each other they don't sting each other. Jellyfish release particular sugars that act like a military uniform. They signal to other jellyfish that they are on the same side and to hold their fire. These sugars are in the formulation too, making your average swimmer chemically resemble a jellyfish. Last, Amit added a lot of calcium. The extra calcium changes the relative concentration of water near the stinging cell, blocking its rush to get inside the capsule. Without that influx of water, the cell can't pressure up enough to deploy.

But does it work? "Yes. Absolutely," Amit told me with complete confidence. To prove his point, on his computer he clicked to a recently filmed segment from a popular Spanish television show called *El Hormiguero*, The Anthill. The show is a mash-up of comedy, interviews, and science, and in this particular episode one of the guest scientists, *El Hombre de Negro*, or The Man in Black, would attempt a daring stunt. I watched a tall Spaniard with a mop of curly hair and a soul patch, who looked a bit like *Entourage*'s Adrian Grenier, stride across the stage, his costume a floor-length black leather cape and black sunglasses. I heard him declare *"la playa"* (the beach) and *"las medusas."* I couldn't understand the fast Spanish, but I interpreted his gestures to mean that he planned to swim with stinging jellyfish. *"No, Hombre de Negro, no,"* said the show's host. Well, this superhero was not taking any chances, Amit told me. The show had previously gotten in touch with the Safe Sea team, who were backstage and had slathered the Man in Black in sting-blocking lotion.

On the show, a wall parted dramatically and the Man in Black strode over to an aquarium containing several barrel jellyfish, bulbous white animals with a piping of purple around the edge. I hear the Man in Black say, *"Muy utricante."* He was exaggerating a bit. These jellyfish, which have become very numerous in the Mediterranean, are considered moderate stingers. Still, I wouldn't volunteer to be in close quarters with one. The Man in Black shed his cape with a flourish and extended his bare arms toward the host for a douse of the sting block. Then the Man in Black stepped onto a trapeze, which lifted him high over the tank. The music boomed. The tension mounted.

"Were you nervous it wasn't going to work?" I asked Amit.

"No, I knew it would. We've tested it a lot. And he had a lot of sting block on before the filming even started."

Amit's product has been through clinical trials in Belize, Florida, and California with box jellies and stinging nettles. In the case of the stinging nettle, only one person felt the pain, compared with everyone in the control group. For the box jelly, two people felt pain, compared with three-fourths in the placebo group.

I watched as the Man in Black was lowered into the tank. It was a little too small for him to submerge completely, so he lost a bit of his coolness by squatting among the jellyfish. He gently teased one of the medusae into his hands. *"¡Qué horror!"* said the show's host. Then the Man in Black passed his bare arm among the jellyfish's oral arms, where stinging cells are most highly concentrated. He was calm, showing no signs of distress.

"¡Brava! ¡Brava!" The audience applauded.

"¡Increíble!" the host shouted.

I heard the Man in Black say, *"Nada."* He felt nothing. He was playing with the jellies now, petting them delicately as if they were kittens.

I decided to spend the night in Tiberias rather than risk a nighttime drive back to Jerusalem with my wonky GPS, and Amit and Tamar graciously invited me to join them for dinner. On the patio of a restaurant overlooking the Jordan River, dense foliage filling the opposite bank, we talked more about jellyfish. They asked me the same question I had asked myself so many times. How did I get here? How did I become fascinated by jellyfish? I told my story about diving in Eilat. About being overwhelmed by the beauty and complexity of the reef.

"You know, that coral reef was said to be made by Medusa," Amit said. Plates of food began to arrive: lemony soft cheese called labneh, falafel, eggplant, shrimp, pizza with artichokes. Tamar ordered and then filled my wineglass with a crisp Israeli white. As day turned to night and the air cooled, a fog wafted off the river beside us, itself so full of ancient lore.

"The mythical Medusa?" I asked. I knew about Medusa's rape, transformation into a Gorgon, and her banishment, but didn't know she had anything to do with coral reefs. Amit, who said he loved ancient stories, leaned back in his chair and told me the end of the myth.

As we know, after Perseus slew Medusa, he cut off her head. He needed the gruesome trophy to show King Polydectes, who had promised to stop harassing Perseus's mother in exchange for proof of the Gorgon's death. When Perseus arrived at home—after a short detour to Ethiopia, where he saved the beautiful Andromeda from the sea serpent—he found his mother cowering, hiding from evil King Polydectes' violent approaches. Perseus pulled Medusa's head from his magical bag and held her astonishing face before the king. Medusa's great powers survived her death. The king turned instantly to stone. In an awful bit of symmetry, a rape victim's head was used to stop another sexual assault.

And then Amit added a rarely told detail of Medusa's tale. After he had decapitated her and before Perseus could properly store the head in his magic backpack, some of Medusa's blood leaked out into the ocean. Just as her stare could turn a man to stone, her blood could petrify too. On contact with the seaweed fronds, Medusa's blood hardened the plants, forming the glorious and intricate coral reefs of the Red Sea. And after my nephew's bar mitzvah, those reefs were where I was headed.

In Medusa's Blood

The day of my nephew's bar mitzvah was hot and bright. My extended family walked together through the winding ancient streets until we reached the remains of the Western Wall of the Temple that stood in Jerusalem over two thousand years ago. We gathered around a small table beneath the blazing sun and listened to the chanting of ancient words in a historic and sacred space. Around us, huge chunks of limestone, deposited there when the Temple was destroyed in 70 CE, reminded us that amid magnificence, there is also ruin. After the service, we wrote wishes on tiny scraps of paper, which we folded over and over until they were small enough to be wedged into the cracks in the Wall. I had done the same thing when I had come to this spot decades before, and I had watched Japanese tourists perform similar acts at the Shinto gate on the sacred island of Miyajima. But this time, I was not a lonely, disoriented student. And as Keith and I helped Isy and Ben search for a spot in the limestone bricks that could contain their dreams, I felt a moment of shared hope.

The next day, our family departed on a car trip south to Eilat, the place where I first put my head in the sea nearly thirty years earlier. We pulled into our hotel after dark, and I could smell the salty air and hear the squawk of seabirds as we unloaded our suitcases. I didn't know whether the reef would live up to my exuberant memories, but I was ready to find out.

In the morning, I woke to the same breathtaking view of the craggy red Jordanian mountains that I remembered. Off to the south, the range continued into Saudi Arabia. On my side of the sea, in Israel, I could see Egypt to the south. Four countries so often at odds shared this little triangle of ocean. And what an ocean. The sparkling blue water was striated in gentle swells and looked like a watercolor artist had swiped indigo and baby blue across the sea's aquamarine surface. Near the shore, the ocean was edged in greenish brown that I knew was a sign of the coral reef below.

I would have gone snorkeling first thing, but I knew that our best chances of a happy dive meant full bellies, so we gorged ourselves at the buffet that came with our hotel room, a cornucopia of fruits, cheeses, yogurts, and baked goods. Then, our rented snorkels, masks, and fins from the beach shop in hand, we crossed the street to the northern edge of the coral reef.

Landlocked as we are in Austin, it was Ben's and Isy's first time snorkeling. They were nervous. Isy put her snorkel and mask on, but couldn't figure out how to breathe through it. "This doesn't work."

"Yes, it does," I said. "You just have to breathe out of your mouth."

"I can't."

"You can. Try two breaths." I demonstrated. She couldn't do it.

"Okay, let's just use the snorkel." I showed her again. "Let's do it together, ten breaths." She got breathing through the snorkel. Then through the snorkel with the mask on. Finally, sitting on the sand, she put her face in, and breathed five times. Then ten. It was slow going.

Ben had more success, but as he and Keith followed a fish a few yards offshore, the top of Ben's snorkel submerged. He sucked in water and came up coughing. When he put his face back in, his mask flooded. Salt burned his eyes. He returned to the beach, pulled his mask off, and flung it in the sand. The reef was tantalizingly close, but after half an hour, our whole family was back sitting on the beach.

"This is going well," Keith said, calmly propping his mask up on his forehead.

I snorted my frustration through my snorkel, which was still in my mouth.

"I'll stay here with the kids. You go ahead." He nodded toward the reef.

I accepted his gift. Submerging in the water, especially with a snorkel and mask on, always has the immediate effect of dampening the clutter in my mind and sharpening my awareness. The soundscape changes entirely, but it's by no means quiet. People talk about the silent sea, but for me the water is noisy. Sound travels about four times faster in water than air, and the reef's soundtrack is the clattering of the sand against gravel on the sea bottom, my own breathing in my chest, and even the grinding sounds of fish chewing. I am aware of the viscosity of water—different from that of air, thicker—and the sensory clues I'm used to receiving from the hairs on my skin are muffled and arrested. On land, I chill easily, but in the water, maybe because of my body heating the thin layer of water glued near my skin, I feel more insulated against the cold. The touch sensors that orient me in the world and that are so beholden to the pull of gravity on land are offset by the buoyancy of water and so are forced to tune in to a softer signal. Because of the mask, my visual range is limited to what's in front of me. Without peripheral vision, I am more predator-like, narrowed to the cylindrical space that ends at the edge of my eyes. My mouth tastes the savory sea as I clamp my teeth around the snorkel. My heart slows. My breathing relaxes. The combined effects of the water literally transform all of my senses and body systems. My brain tries to recalibrate and make sense of these changes by connecting to something more primal, simpler, and more elegant.

The coral reef was protected by a rope held in place by buoys, but the reef edge was easily accessible, and snorkelers dotted its fringe like hikers along a trail. Looking downward as I kicked toward the reef, I began to see small mounds of coral cobbled together in the sand, colonizers sent out scouting for new habitat from the mother reef. With the constant grinding of their jaws, parrot fish, looking as if Isy had colored them using pastel sidewalk chalk, worked hard at returning these upstarts back into the surrounding sand. Closer to the reef, the coral mounds grew larger and more established. I dove down to explore. Tiny orange fish darted in and out of the caves made by coral ledges. Thick-fingered fans emerged from nooks and black curved fans dotted the crannies. Lichen-green brain coral squashed against blue stubby columns and the orange handlike projections of fire coral. Narrow fish in black and white stripes, black fish in a single white stripe, white fish in black polka dots swam through my field of view.

When I've had the chance to dive on coral reefs in the past decade, which hasn't been that often, I've emerged with a sick feeling at seeing more rubble than living coral, mucus strands splayed across dying colonies, and algae coating the skeletons. I expected to see the same decline in Eilat, but the vibrant reef of my memory still existed. It was all here. An enchanted playground of arches and spires filigreed in pastel corals. Abundant fish flitted and flirted with me, guiding me into the passageways of living creatures moored together in collective architecture. I was back in 1987, aswirl in paradise.

But then a rumble broke through my reverie, through the ambient music of the ocean. And it was angry-sounding. Someone was bellowing through a loudspeaker in Hebrew. I lifted my head up along with a dozen other snorkelers. Their masked eyes were turned toward me. I spun in a slow circle, taking in the shoreline and then the rope. I was on the other side of it. I was the problem. I'd trespassed into the marine protected area, where swimming was prohibited. *Slicha! Slicha ma-ode!* I apologized in Hebrew and then in English, not that anyone could hear me. Sorry. Very, very sorry. I swam back out of the protected zone as fast as I could.

In the afternoon, I stood in front of the Interuniversity Institute for Marine Sciences in Eilat, which looked like a military outpost, a nondescript concrete building secured with a heavy metal fence. It was just as I remembered it. Standing outside, I called the director of the institute, Maoz Fine, who opened the imposing gate. He was fit and tall, with a close-cropped head of dark hair, and he spoke in a manner that was both soft and confident. Maoz was a leader in the science of coral reefs, especially trying to answer questions about the future of corals, a future that is in great jeopardy. While some jellyfish may have a future ripe with possibility, their close cousins the coral are in peril.

Maoz and I walked toward what I quickly remembered as the classrooms where I first learned the melodic Latin names of twenty different coral species. Now it was a kitchen for the laboratories with a million-dollar view of the red Jordanian mountains and the rich blue sea. Near shore, a dock with two research vessels bobbing at their ropes hovered over the brownish water lapping at the grainy shore. The brown was not a sign of pollution but a signal of the coral reefs below. We turned away from the sea and walked toward a concrete slab on which several rows of small aquaria were lined up. Overhead a roof of netting let in the changing sky conditions but filtered the strength of the semitropical sun. A finger-size piece of coral rested on the bottom of each aquarium. After almost thirty years, the genus name still popped easily into my head. Called *Stylophora*, it was one of the most common on the coral reef.

Maoz named this outdoor lab the Red Sea Simulator because it allowed him to simulate conditions of many different future Red Seas. The seawater in each of the aquaria was pumped from thirty feet below the surface to preparation tanks. Maoz then altered the acidity and temperature in each of the aquaria to match future conditions expected from climate change. Something that looked like a boxy R2-D2 from *Star Wars* propelled itself down the aisle between rows of aquaria, buzzing and clicking. It carried a horizontal bar stretching the width of the aisle. On each end of the bar, twin sets of instruments pivoted downward, dipping below the surface of the water into two aquaria across the aisle from each other. After a minute the instruments pivoted up out of the water. The robot rolled to the next set of aquaria, and the instruments pivoted down, testing the water again.

Maoz had built this robot to automatically and regularly sample the conditions in the test aquaria. The information was wired to computers in his lab so that he could keep constant tabs on what was going on in his many Red Seas.

When water temperatures rise by about 1 degree Celsius, or about 1.8 degrees Fahrenheit, the algae that live in coral's tissues and pay rent to their host in the form of sugar produced by photosynthesis bang on the thermostat to see what's wrong. Finding no fix to the heat problem, the algae terminate their lease and move out. The corals lose their healthy greenish glow from the algae in their tissue and become bone-white. Without the sugar previously collected from their algal tenants, corals wither. The process is known as bleaching.

For today's corals, bleaching is the number-one cause of death. In 1998, extremely warm temperatures damaged corals in fifty countries and killed 16 percent of the world's reefs. In 2010 warm waters again caused global bleaching events. In 2016 in the northern region of the Great Barrier Reef, 67 percent of the corals bleached and died, the most ever in one year. The International Society for Reef Studies has reported that since the 1980s a third to a half of all coral reefs have died. They predict that by mid-century, most coral reefs—home to more species than any other ecosystem in the sea—will no longer exist. The concomitant loss of the billions of fish and invertebrates that make their homes on the coral is a heartbreaking side effect. But here's a surprise: More than thirty years after the first reports of coral mass-bleaching, the corals in Eilat have not bleached—even when temperatures rose more than 2 degrees Celsius, as they did in 2010 and 2012. Maoz and his collaborators think there's something special about these corals.

To understand what, they incubated five species of corals in the Red Sea Simulator for a month, slowly turning up the heat on the corals until they were bathed in water 6 degrees Celsius, or 11 degrees Fahrenheit, warmer than normal—six times higher than the temperature that normally causes corals to bleach. The results? Nothing happened. The corals could handle the heat. The algae didn't move out. The corals didn't bleach. Only when the temperature was boosted to a simmering 7 degrees Celsius higher than average did the number of symbionts drop, and then only by a fraction: 30 to 15 percent, depending on the species. "Stressing the corals here is difficult," Maoz told me. "They are very unique in that we find resilience to temperature increases in all our experiments."

Maoz and his colleagues think that the Eilat corals are so heat tolerant because of the way their ancient history is tied to the geography of the region. The Red Sea is long and thin, connected to the Indian Ocean by a shallow strait located off the coast of Egypt that's just over 100 meters deep. During the last glacial period, a lot of seawater was bound up in ice and the sea level dropped by 120 meters, making that strait into a dam and essentially turning the Red Sea into the Red Salty Lake. Evaporation coupled with the lack of water exchange with the larger sea made the lake saltier and saltier, and the corals living there died. When the glaciers melted about eight thousand years ago, the water overtopped the dam, and the Red Sea was once again connected to the Indian Ocean. Coral larvae from the Indian Ocean flowed toward the vacant reefs in Eilat. But before they reached this unclaimed turf, they had to pass over that shallow strait, where the sun beat down, warming the water. Only the most heat-hardy propagules—the fertilized eggs or larvae or even bits of broken coral—made it through the temperature barrier. Maoz calculated that the bleaching threshold for the Eilat corals is just about the same as the temperature barrier that the larvae had to pass through, about 32 degrees Celsius, or 90 degrees Fahrenheit. At present, the average water temperature around Eilat is 27 degrees Celsius, or 80 degrees Fahrenheit. Those few degrees buy the Eilat corals some time. In 2050, when most other corals in the world will have crossed their bleaching threshold, the Eilat corals will still have 4 full degrees to go. It'll be over a century before summer temperatures are hot enough for a mass bleaching in Eilat. "These are likely the last coral reefs that will survive global climate change," Maoz said.

But a future ocean won't just be hotter; it will be more acidic too. To understand the impacts of lower pH of the ocean, climate change's evil twin, Maoz grew corals in an acid bath about 1 pH unit lower than today's seas, which corresponds to ten times more acidity. After a month, the acid changed the creatures fundamentally: The corals lost their skeletons. The polyps elongated and tripled in weight. Individuals separated from the colony, forming long skinny tubes with a crown of delicate tentacles surrounding their mouths. In this floppy form, most showed no evidence of bleaching but kept the deep green color of their algae. By the end of the year, these animals didn't look anything like the hard, stony corals that we think of when we think of coral reefs. They looked like sea anemones.

Because corals have skeletons, their remains are preserved in layers on the seafloor. Scientists can drill a core through those layers and examine it to look back at history. Just as thin bands of tree rings indicate years of drought, thin layers of coral skeleton indicate the decline of coral growth. These eras are called reef gaps. What's interesting about these gaps is that, though the corals disappear, they reappear too. Were Maoz's floppy creatures evidence that corals survive rough periods by going soft? It's possible. After a year in the acidified water, Maoz put the skeleton-free corals into rehab. He returned them to water with the same acidity as today's seas. The same creatures started rebuilding skeletons. They joined back together into colonies. They became reef-building corals again. However, just because the corals survived acidification doesn't mean that it isn't a problem. During periods when corals stop building reefs, everything that lives on the reef is suddenly homeless. Coral skeletons are the infrastructure that allows for the great biological diversity of the reef.

Besides the chance to walk memory lane back to Eilat, Maoz's ocean acidification work was one of the major reasons I wanted to visit this lab. I wanted an answer to the question that got me started on the jellyfish odyssey in the first place. Would jellyfish really be winners in a future acidified sea? In the years that I'd been following jellyfish research, I had found only a few studies that touched on the question.

"I don't like the assumption that ocean acidification won't affect them,"

Maoz told me when I asked him about the fate of jellyfish. "They are highly dependent on their statoliths." Statoliths are the tiny grains of gypsum in the statocyst that tell jellyfish which way is up and which way is down. "Gypsum already dissolves in today's seawater, so they are sensitive. We did a study . . ." At this point we were talking in his office, and Maoz pulled out a piece of paper from a stack on his desk and pointed. It was a list numbered 1 to 10 with Hebrew writing next to each line, which I couldn't read. "It's on my to-do list to get it written up." Here was an answer to the question I'd wondered about almost five years earlier, on someone's to-do list halfway across the world.

"What did you find?" I asked.

Maoz told me he'd studied two different species of jellyfish. "We raised them under acidified conditions and then examined their statoliths. We also extracted the crystals from their statoliths. The statoliths were smaller and the crystals were shaped differently from normal. The life cycle was affected as well. It wasn't a big study, but it showed that jellyfish are not insensitive to ocean acidification."

"How did the jellies swim in the acidified water?" I wondered.

"They were sluggish. Like they were drugged. The biggest risk for jellies and low pH is definitely sensory." I thought about another cohort of jellies that suffered from malformed statoliths, the jellies that Dorothy Spangenberg sent up on the *Columbia* space shuttle. Like those jellynauts, the jellies of the future would be on another trip of our making, an acid trip. But it would take place right here on our own planet, and like the signature drug of the 1960s, this acid could also mess up their senses. A few other studies of jellyfish exposed to acidic conditions have now been published. While one study from Spain found that jellies could withstand the stress of acidified seas, scientists in Japan and England found similar results to those that Maoz described to me: The jellyfish they studied swam strangely in low pH waters. We may find that some jellies in some places are better at surviving acid trips than others.

Maoz continued: "With coral, we can see how they survive ocean acidification. Their metabolism has to change. There are changes in tissue formation. And we know some species have survived past acidification events. But how do jellies do it? It's not in the geologic record. Do jellyfish find an oceanographic refuge? A place in the ocean where the ocean acidification isn't so bad? Is it behavioral? Or do they change their life cycle, hunker down as polyps or just hang out in the surface as medusae? The jellyfish have a lot of choices, but the fossil record doesn't tell us anything about their life history." There is no historical record of jellyfish to drill through the way you can drill through the remains of ancient coral reefs. Jellies don't have skeletons and they aren't stuck in place. They are literally much harder to pin down.

Before I left, Maoz mentioned that he'd been swimming off of Tel Aviv and that there was a big bloom of the nomadic jellyfish. That was the bloom I had read about on my Google alerts and was hoping to see, so I asked for some details about where and when he saw them. I would be in Tel Aviv in two days. I hoped they would stick around.

After a miserable morning not getting the hang of snorkeling, Ben and Isy weren't itching to try it again. But I wanted them to experience the magic that had pushed me to study the sea, that had brought so many conversations about the ocean to our inland dinner table, and that had overshadowed our lives for the last half-dozen years.

When I was at Safe Sea, Amit Lotan had told me about a shop that offered guided snuba dives, a cross between scuba and snorkeling. Before Jacques Cousteau popularized scuba, another Frenchman, Maurice Fernez, had developed a mouthpiece that could feed a diver air through a long tube from a canister on a boat at the surface. Snuba brought back that technology: no training required, no diving up and down, no chance to get the bends from compressed air, no tanks to lug around. It seemed the ideal way to share the reef with the kids.

The snuba center was located near the southern half of the coral reserve.

It was nothing more than a beach cabana with hammocks strung off to one side. Makeshift tarp patios furnished with old car seats littered the front of the joint, as if people went on extended snuba campouts. A few divers sat around on mismatched furniture half wet-suited up.

"This looks sketchy," Ben observed as we pulled up in our rental car.

"It'll be fine," I replied, hopping out. Dive shops tend to come in two styles: edgy surfer or sleek and polished. This just happened to be the former, and frankly I felt more comfortable with divers who weren't overconcerned about the polish. It meant there was plenty of time to think about safety. Still, I could tell by the way they shuffled: Neither kid was fully convinced that this was a good idea.

Our guide was a serious woman named Natalie, who introduced herself without cracking a smile. She gave us the rundown on snuba, demonstrating how the regulator worked and explaining about keeping a calm body underwater. Even though we would be staying shallow, limited by the length of our hose, we would have to equalize the pressure in our ears by swallowing or by holding our noses and blowing outward. Both kids were glued to her every word and obediently practiced equalizing. Natalie made sure all our masks fit properly, that our wetsuits were comfortable, and that we were wearing the proper weights. Her low-key demeanor helped calm the kids' nerves.

We walked across the street to the beach. Scuba tanks were already loaded into small inflatable boats that would carry our air. Natalie handed each of us a regulator attached to the long hoses. She explained sternly that the reef was a marine protected area. We were to touch the seafloor only in areas designated by ropes. Outside those places, we could look but not touch. I sat down, put on my fins, put the regulator in my mouth, adjusted my mask, and descended, popping open my ears as I went. By the time I looked back, everyone had followed successfully. I could see the dancing smile in Isy's lavender eyes. Keith looked thrilled. Ben even waved.

We were diving in a part of the reserve called the Caves, and the coral reef was as lovely there as in any place I'd seen. With the luxury of time underwater, I began to recognize many of the coral species I'd known. Like old friends, their names returned: *Millepora*, the fire coral, which burns if you rub against it. *Acropora* and *Stylophora*, branching corals. *Fungia*, a coral formed from a single polyp that pools around itself, looking like a mushroom. *Pocillopora*, little round buttons of coral. Isy pointed out a vibrant giant clam, its green speckled mantle spread wide. Blue feather dusters, the tops of worms buried in corals. Broad lavender sea fans waving gently in the surge. Ben stared down a school of orangey-red squirrel fish with big black eyes that scurried deep into nooks when he approached. Keith trailed behind a plump parrot fish munching away. Isy noticed an orange-and-white Nemo hunkered down in its anemone. Long blue coronet fish—those sprinters that are making their way across the Mediterranean—hunted in groups and alone. A brown-striped lionfish sailed by, its fins splayed.

The dive ended much too quickly. Natalie signaled, and we followed her back to shore between the ropes. Climbing out of the water, the first words out of Ben's mouth were not what I expected, but what I'd hoped for. "It's over?" And then, "It was so peaceful." Delight rippled through my soul. It had worked. Experiencing the ocean's brilliance firsthand meant more than all my after-school lectures combined. It meant more than any nature video or TV show he'd ever seen. And it made this crazy jellyfish adventure worthwhile. Our kids need to understand what's at stake too.

Two days later, I was walking through the gates at Tel Aviv University, early for my meeting with Yossi Loya, the professor who had taught my first marine biology course in Eilat. I hadn't eaten, so when I passed by a café on the ground floor of one of the concrete structures, I headed inside. They served pita, my old nemesis. I ordered one, and it arrived oozing with hummus and studded with bright green falafel. It was horribly delicious.

As I exited the elevator to the floor where Yossi has his office, I passed an advertisement for marine-ecology classes showing divers on the Red Sea reefs. Besides being glossy and full-color, it wasn't so different from the paper sign I had fallen for all those years ago. I turned away from the memory and recognized my first marine biology teacher. Now in his seventies, Yossi is still energetic, with a broad, tanned face and the thick shock of white hair I remembered from thirty years ago. Yossi had graduated from the lab of one of the world's most notable ecologists, Lawrence Slobodkin, at the State University of New York in Stony Brook. Studying the rocky beaches of Long Island, Slobodkin had developed some of the first mathematical equations used to calculate species diversity. When he returned to Israel, Yossi thought he could apply some of Slobodkin's ideas to the coral reefs in Eilat. The conditions were primitive, Yossi told me. The marine station had just opened, and it was nothing more than a cinder-block building on the beach. "But once you put your head in the Red Sea, you are doomed," he said.

Didn't I know it? Ben's awed reaction to the dive on the coral reef was still fresh in my mind.

The quantitative tools that Yossi developed to measure coral diversity in those early years have now been used by scientists the world over. In Japan, Africa, Australia, and the Caribbean, researchers assess coral health and vulnerability to environmental problems like oil spills, coastal development, and climate change using Yossi's methods. Yossi himself has used those tools to measure the impacts of a string of threats to the coral in Eilat, then to fight for the health of the reef. In the 1970s, huge supertankers full of oil drove into the Gulf of Eilat and offloaded at a pier just north of the coral reefs. Between 1972 and 1980, there were three or four oil spills in Eilat per month, killing the coral on contact and affecting the reproduction of those it didn't touch. Between 1986 and 1996, the town of Eilat grew from a sleepy seaport of twenty thousand inhabitants to a booming beach resort of about fifty thousand. The town's infrastructure hadn't kept up; over 700 million gallons of raw effluent flowed directly into the sea each year. "You could see a river of sewage running directly into the Gulf," Yossi said. "We screamed, 'What's happening to this jewel of our shores?'"

After a long political struggle, Eilat built a modern sewage treatment plant located seven kilometers north of the city. But the reef's troubles still weren't over. Next, two commercial fish farms in the northern part of the Gulf of Eilat ramped up production by a factor of ten. Pollution in the forms of fish excrement and fish food that washed out of the cages poured into the water, smothering the corals. The battle to evict the fish farms was long and contentious. Yossi said, "I won't go into the dirty stories, but I had threats on my life."

"But when I was on the reef the other day, it looked beautiful. Was I imagining that?" I asked.

Yossi showed me a graph proving that my memory wasn't so bad. The first time I dived in Eilat was 1987, following the period of oil spills but predating the problems with sewage. At that time, the coral covered half the reef, and in a ten-meter stretch of reef (about thirty feet) you could count an average of fifteen different species of coral. The years between 1987 and 2006 nearly obliterated the reef. At the worst, during the time of the fish farms, 80 percent of the reef was dead. A ten-meter stretch would include only nine different species, a decline of 40 percent. But now the reef was back to the same condition it was when I first saw it, maybe a little healthier. Coral cover was over 50 percent, and if I had counted, I could've expected to see eighteen different species in a ten-meter swath. The species diversity was even greater than the last time I was here. The thirty years between my dives in Eilat had been filled with destruction, but by avoiding Israel, I had skipped right over it.

The health of the Eilat reef is due not just to improved infrastructure, but also to environmental stewardship and regulation. Much tighter conservation measures are in place now than in the past. As I learned by breaking the rules, access is highly restricted in the Marine Preserve. Divers are allowed access only at specific entry points, and the park is patrolled. Commercial fishing is forbidden and nutrient inputs are managed through a sewage system. Threats exist, of course. There's still an oil spill roughly every year. And new plans for fish farms continue to be discussed, which could compromise the clean water. But my return to Israel taught me that this reef, one that has always felt so necessary to me personally, may in fact be the last coral reef on our planet. In the most unexpected place, against the terrible odds of climate change, political turmoil, and an ever-growing coastal population, I found a little corner of the world, at least, where our seas were being saved. Its story is a symbol of how we can decide to preserve the precious gifts of the sea.

I returned to our hotel just as Keith and the kids returned from spending a couple of hours at the beach with my sister and her family, who had met us in Tel Aviv after we returned from Eilat.

Dusting off the sand from her ankles, I asked Isy, "Any jellyfish?"

"No," she answered. "Want to see the dance that we made up?"

"Are you sure? Were there any purple flags flying? Was anyone talking about jellyfish?" I pushed.

"No, Mom. Can I use your phone to play the song?"

"Because I heard there was a bloom at Tel Aviv beach," I insisted.

"Mom, there weren't any jellyfish."

"Okay," I said, handing over my phone. My time was running out. We would leave in two days.

21

Party Like a Jellyfish

During the bulk of my stay in Israel, Bella Galil had been working in Italy, digging through dusty archives for mentions of animals native to the Mediterranean so she could compare them with what exists in that sea today. She returned on my last full day in the country, and we planned to meet for dinner in Tel Aviv.

When I arrived at the restaurant, the sun was setting, casting its long pink glow through the windows. I greeted Bella, and as I sat next to her on a couch with a fabulous view of the sunset, the waiter handed me a large menu. I broke the ice by mentioning that for my entire trip I'd been on the hunt for the nomadic jellyfish bloom and had failed to find it.

Earlier in the day, on a tip from Facebook, I'd driven up to Caesarea, an hour north of Tel Aviv, to try to find the jellyfish bloom. When I arrived, loads of people were swimming, a sure sign that I'd missed the jellyfish again. I walked down the sand, past two sunglassed teenagers gossiping on towels and a cluster of smaller kids wading in the shallows. Yellow, red, and blue windsurfers flitted across the water like giant butterfly wings. Young boys were digging shallow holes in the sand. One pulled up the thin remains of something gelatinous and wiggled it in his friend's face. Then the boy wrapped the gelatinous remains around his wrist like a watch and showed it off to his friend, a truly timeless accessory. It was as close as I had gotten to seeing a jellyfish since I'd come to Israel.

"Why didn't you ask me?" Bella exclaimed. "I always know where the jellyfish are. I just heard today that the bloom is in Haifa Bay. It's a huge bloom, by Kiryat Yam."

I scribbled "Kiryat Yam" in my notebook even though I didn't expect to get there. "Too bad. I leave tomorrow."

Bella tilted her head and lifted her eyebrows in an Israeli way that said, *Your choice.*

Bella was an elegant woman in a well-tailored dress and pumps; she didn't look like a field-worn biologist. It wasn't that Bella hadn't done her share of fieldwork; it's just that as her career progressed, Bella understood that fieldwork could take her only so far. The realization that scientists need to roll down their sleeves and button up their jackets has dawned on the European scientists I met much more insistently than it has on those on the other side of the Atlantic. When I was a graduate student, I felt an unwritten but clear line between science and policy—and for good reason. Scientists need to draw conclusions from data. Their results should not be biased by industry or politics. But the division between what scientists discover and what they think needs to be done about it seems to eroding—and again for good reason. In the United States as well as Europe, there is an ongoing attack on science, a willful insertion of misconception into the public sphere. And that leads to policies that fly in the face of data and reason.

After my story about the Suez Canal failed to bring any attention to the scientific predictions about the results of the expansion, I had a heart-to-heart with myself about policy and my role in it. I'd spent many years learning about our planet and what threatens it, and I understood that it is irresponsible to ignore such knowledge. This book was certainly something I could do to increase awareness, but at that point, I had no idea if or when my jellyfish writing would ever get published.

Landlocked as I am in the capital of Texas, I realized that the most important conversations around me were not about the sea—not directly anyway—but about energy. Texas is the battery of the country, the home of the nation's foremost oil and gas industry. Although China is gaining ground, the United States has produced more carbon emissions than any other country. Per capita, it's not even close. Because of its oil and gas industry, Texas produces more carbon dioxide than any other state in the country, double the amount produced by California, which is second. As of 2016, if Texas were a country, it would have the seventh-largest emissions output in the world, just behind Germany. Carbon dioxide in the atmosphere was what got me started on this long and winding jellyfish road in the first place. It produces global warming's evil consequence, ocean acidification. If I was going to help save the oceans, I could try to do it through the atmosphere.

I found a local group of citizen lobbyists who supported a solution to the carbon problem that made a lot of sense to me, one that, rather than burdensome regulation, emphasizes a market-driven correction called carbon fee and dividend. They taught me that there are some ways to use your voice that are more powerful than others. Letters to the editor, which seem such an archaic form of free speech in these Twittering times, actually get read in Washington. If you send a published letter to your representatives with the weight of a newspaper behind it, they pay more attention than if you send out an electronic chirp. If you pick up the phone and call your representatives, they (or their staff or machines) have to take your call, even in political districts as gerrymandered as mine. If you meet with the people who represent you in Congress, you can sit down in a frank discussion about your idea, even if they fundamentally disagree with you, as mine disagree with me. The people elected and paid to represent you don't know your thoughts if you keep them to yourself.

I started doing all these things. You'd be amazed at how easy it is to write to your local paper and get a letter published if you have something reasonable to say. And people read the letters. Every time one of my letters is published, a retired man named Sheldon, who doesn't agree with me at all, writes a physical letter to let me know about it. He's my old-fashioned troll. His letters initially made me nervous, but now each one is a reminder to keep getting my opinion out there. I'll bet that Sheldon never misses a vote or a chance to tell his congresspeople what he thinks. Sheldon reminds me that I'd better keep participating in the discourse if I want to see anything change. Speaking up matters.

Bella, for her part, is very vocal. As our appetizer arrived—an invasive species of shrimp that now makes up the bulk of the commercial catch, served carpaccio on a bed of ice—she told me that she had continued to collect names on her petition about the problems with the expanded Suez. She had now gathered over five hundred. Bella was regularly quoted in newspapers and on television and had remained involved in more United Nations working groups than I could keep straight. Our conversation wandered from her long scientific career—which started at Cornell, where as a student she studied stick insects—to ways of identifying new invasive species and then to what she sees as a real crisis in the Mediterranean.

Buoyed by the crisp white wine and playing devil's advocate, I argued that the Mediterranean might be better off with more Red Sea animals, maybe even coral reefs one day.

She responded that it doesn't work like that. Entire ecosystems aren't imported wholesale. The Mediterranean wouldn't just become the Red Sea. The particular animals that establish themselves in a new place are the opportunistic ones, not the type that slip unobtrusively into the ecosystem. In order to establish populations, these invaders must reproduce quickly and have few or no predators.

But what about marine protected areas? I asked. The Mediterranean has a strong system of MPAs. Don't they act as a sanctuary for native species?

Not as much as you'd think, she replied. In fact, she said that marine protected areas were being taken over by non-indigenous species. "What we have in the Mediterranean is a network of MPAs protecting invasive species. Nature didn't read your article about the Suez Canal. The sea doesn't know that it's going to be inundated with new species. The animals keep coming and the opportunistic ones are the ones that remain."

This was terrible to hear. I had seen MPAs as the way to protect the oceans, a silver bullet, but of course that was an oversimplification. While MPAs remain the best tool we have to protect the ocean's health in most places, here was an example where it wasn't sufficient. Just as asking whether jellyfish are increasing globally was the wrong question—what really matters is why they're increasing in the specific places where they are increasing—there is no one-size-fits-all solution to the ocean's problems. Each ecosystem is unique and is subject to different pressures. Taking into account the nuances requires more work and more understanding. Bella was pointing out that if we fail to see the ocean in its glorious complexity, we may fail.

"So what can you do?" I asked, despondent.

"Raise your voice. By staying quiet, we are deceiving the public. The public deserves to know what's going on in the ocean. We are being willfully myopic." Indeed, the one thing unifying the ocean's problems is that fact that what we do on land ultimately has consequences for the sea.

Back in my hotel room, I wrote Bella an e-mail, thanking her for the dinner and the provocative conversation, upsetting as it was. I tossed in a line asking if she could specifically tell me where the jellyfish were in Haifa Bay. The last sentence I wrote in my notes that night was "Do we want to gamble?"

I could tell it was early even as I fumbled for my glasses. My phone said six a.m. I reflexively checked my e-mail. A half hour earlier, at five-thirty, Bella had written back to me, responding to my late-night e-mail. She said the person who told her about the jellyfish was named Moti Mendelson. She included his phone number and wrote that she would try to contact him for me to find out exact details.

It was all the encouragement I needed. I slipped out of bed, grabbed my swimsuit, mask, snorkel, and underwater camera, and stuffed them in a backpack. I took the rental-car keys, kissed Keith, and told him he was going to have to pack up everyone for the airport. I was going to Haifa. I would try to be back by early afternoon, and I would call later. He seemed to nod in assent. On the way out of the hotel, I stopped to grab a cup of coffee for the road, but the café was still closed. Dang. My day never started without caffeine.

The rental car's GPS reported that Kiryat Yam, if I was reading my scrawled notes from the night before correctly, was about an hour-and-a-half drive ahead. Since it was early, there was almost no traffic and I hardly got confused on the roads leaving Tel Aviv.

About forty-five minutes into the drive, my head started to tingle. I needed coffee. But I passed up the next three rest stops, worried about the actual logistics of ordering coffee in Hebrew. North of Caesarea, I could tell that the pounding would start soon. I turned into a rest area. But I chose poorly. It was shuttered. I zoomed back onto the highway; my pounding head would just have to wait.

The highway into Haifa passes through two very long tunnels. They cut underneath the bulk of the city, which, like so many ancient cities, sits on a high promontory with a great view of any invading forces that might approach. Emerging from the tunnels, I encountered a sweeping view of the semicircular bay, which stretches twenty miles. I could barely see the ancient city of Acre perched on the bay's northern shore. I hit my phone's e-mail button. Bella still hadn't written back telling me where exactly on this long shoreline I would find the jelly bloom.

My GPS showed that Kiryat Yam was situated about halfway around the semicircle, on a very long stretch of beach. It also showed that several roads on my left would take me more immediately toward the ocean than the current route my GPS was following. I spontaneously took a left, reasoning that if I drove along the beach road, I might be able to see jellies washed up in the sand.

The street I'd turned down was lined with apartment buildings, five or six stories high. It was a nice neighborhood, not particularly upscale but comfortable. Some people were dressed for work and making their ways to their cars. Others were taking their dogs out for a morning walk. While I felt uncertain, everyone else seemed like they knew what they were doing. Here I was, trying to somehow run into either jellyfish or some dude named Moti Mendelson.

Ahead, the street ended at the beach road. My GPS was frantically re-routing, telling me to turn right. But at the corner, I saw an open parking space. And no parking meter or other signs. I pulled into it. I grabbed my backpack and scrambled through traffic across the beach road. I stopped on a concrete slab, marking a spot where a building once stood, to survey the shoreline. It wasn't beautiful but wasn't the worst I'd seen either. A few scraps of trash and fishing nets were strewn in the murky sand. A couple of fishermen cast lackadaisical lines into the water. Scattered couples and groups of retirees in swimsuits walked the shore, out for some exercise. Beyond the beach, the bay itself was sparkling in the early-morning sun.

Just then, in front of me, a man rose like a sea monster from the water. He stood knee-deep in the gentle surf. A snorkel and mask were perched on his forehead, and he was holding a very big underwater camera with a huge arm that extended above it holding the flash. If anyone would know if there were jellyfish out there, that sea monster would.

I picked my way down a rubble berm and approached him. "Do you speak English?"

"Yes."

Gesturing toward his camera, I asked, "Did you happen to see any jelly-fish out there?"

His eyes knitted together. Was that a weird question? I suddenly wondered, reminding myself that not everyone thought about jellyfish as much as I did. But he answered, "Yes, I'm here to take pictures of the jellyfish."

My heart started pounding. "Are there a lot?"

"Millions."

Wow! Jackpot!

"Were they *Rhopilema*?" I asked, referring to the nomadic jellyfish.

"And *Phyllorhiza* and *Cotylorhiza*." He said he dived in Haifa every day. Today he had come to take pictures of the massive bloom in the early-morning light, but as luck had it, his camera had run out of batteries just when I arrived.

Something was dawning on me. I knew the name of only one person in all of Haifa, but that someone knew about jellyfish. I pulled out my phone and double checked Bella's e-mail. "Are you . . . Moti Mendelson?"

His eyebrows shot up and he took a step back. "You know me?"

"No way!" I hooted. I explained about my dinner with Bella Galil the night before, when she told me that someone named Moti reported the bloom. I had only a few hours left in the country, so I'd hopped in my car, hoping to find him. Against all expectations, in this particular spot, on this long stretch of beach, here he was.

Moti started to walk away from me. Had I offended him? Had the coincidence freaked him out?

He turned and said, "Come on. We need to go buy a lottery ticket! It's a lucky day." He was right. If anything had gone differently that morning, this meeting wouldn't have happened—if I'd woken up earlier or later, if I'd stopped for coffee, if I'd followed my GPS instead of making a random left turn, if I'd not found a parking spot, if Moti's camera had kept working. If any of those things had happened differently, I wouldn't be standing in this spot at this moment. Life may be a series of happy accidents, but I had never before had them all come together in one morning.

Moti was not a young man, but he was incredibly spry. He had a gray mop of hair and a tightly trimmed beard. Later—over a very long-awaited cup of coffee—Moti told me that he hadn't had much of a childhood. He'd grown up in Jerusalem but didn't have a stable home. It seemed that the army had taught him more than anyone else in his life. That made him tough and a little arrogant. But he was also smart, and the combination made him a little crazy. He said that he always expected he'd get less crazy as he aged, but it hadn't happened. When I asked Moti if the jellyfish had been stinging him, he looked down at the welts crisscrossing his legs and forearms and said, "Yes, but I like the sting. It keeps me healthy. If I don't get stung at least once a week, I find a bee and make it sting me." He qualified: "Not one of the African bees—I'm not that crazy—a European one."

Moti asked me if I was planning to go snorkeling.

"Yes," I said, gesturing to my mask and snorkel in my backpack.

He shook his head. "It's not a good idea. The jellyfish are starting to die and the water is full of tentacles. Just offshore there is a wall of jellyfish. If you go to the wall, you will not be going to the airport tonight; you will be going to the hospital."

Going to the hospital was not an option for me. But neither was missing out on seeing this bloom of jellyfish that I'd finally managed to find. Moti could see I wasn't backing away from it, and offered to be my guide in the water, to take me to parts of the bay where I could see jellies safely.

Moti told me to wait for him while he went home to get a different camera. In the meantime, I pulled out my bottle of sting block and applied layers to every bare spot on my arms, legs, and face, hoping that it would work as billed. When Moti returned, I offered him the bottle. He shook his head. Oh, right, he liked the sting.

We put on our masks and walked into the water. It wasn't long before we started to see jellyfish rolling on the seafloor. These medusae were dying or already dead. They ranged in size from grapefruit to cantaloupe. Some were bulbous and others were more saucerlike. They were white or a pale, icy blue.

The water was full of tentacles. I could see strands floating all around me. Trying to avoid them was impossible. The sting-blocking lotion seemed to be working, though. I wasn't feeling pain, but I watched Moti slap at his exposed arms and legs as he was stung.

Moti lifted his head from the water. He told me that there was wreckage from an old bridge in the bay and that was the marker we were headed for. I followed his lead.

We started to see more jellies, some swimming near the surface of the water. Moti pointed to a beautiful animal the size of a volleyball. Its eyelash-thin tentacles formed a delicate fringe around its bell. Below, the four oral arms terminated in thick pyramids that looked cut from a fine-filigreed crystal. Moti continued to lead us onward. More and more jellies were alive and swimming around us and below us. Moti careened about like a dolphin, playing with the jellies as though they were living toys. He reminded me of YouTube videos I'd seen of dolphins playing with jellyfish. Moti knew that the stinging cells are localized in the tentacles and the undersides, so he caressed the jellies on the tops of their bells. He rolled them over gently and passed them between his hands delicately. I gingerly touched the top of one animal. It felt warm, but then the ocean was warm too. There was a firmness to it, like vinyl. Very small bumps covered the surface. Moti poked his head above water. "I love them so much. They're like dancers." He was right. The fluid movements recalled ballerinas in elegant white tutus.

As we swam above the wreckage of the bridge, the number of medusae increased. We started to see lots of fish too. They were small, about five or six inches long, and sported handsome stripes. Moti told me they were originally from the Red Sea and had become established in this part of the Mediterranean. Below us a group of fish swarmed around the jellies rolling on the seafloor, apparently unaffected by their stinging cells, protected by their own version of sting block. A small school had excavated a fairly large hole in the top of one dying medusa. The fish were ducking in and out of the excavation, grabbing a bit of jelly flesh with each visit, like partiers visiting a buffet. They were feasting on the jellyfish's gonads, the most nutritious part, which are located near the top of the bell.

A moment later, *zap!* I was stung across the front of my left thigh. It was a sharp, scalding pain. I looked down to find the tentacle but saw nothing obvious. The pain was strong, but it didn't intensify. It felt like a wide band of wasp sting. It hurt, but I could stand it. Before we were through, I would be stung a few more times, along my ankle and the back of my knee, places I probably missed with the sting block or where it had worn off.

As we swam on, I noticed a jelly that looked different. Instead of a smooth, round pale bell, it was flatter and polka-dotted. It was a loner, a different species, caught up in the dance of the nomadic jellyfish. Moti dived downward and picked it up. As he swam to the surface, he placed the spotted jellyfish on the crown of his head. Lifting his face above the surface, the jelly still balanced on top like a yarmulke, he said, "I could wear this in Jerusalem, at the Western Wall." I laughed through my snorkel.

To keep us away from the wall of jellies, Moti kept gauging our distance from the shore and in relation to the wreckage of the bridge, but we must have been drawing closer, because the jellyfish were growing thicker. Around us, strands of tentacles that had been ripped off by shear currents or perhaps nipped off by fish glinted in the sunlight like pieces of confetti. Medusae decorated the sea surface like charmed chandeliers sprung to life. Around us, the pale ballerinas swirled and floated in their frilled skirts. Below, senescent creatures rolled gently in the surf like balloons that had fallen to the dance floor. It was as if the ocean had thrown a massive party. But the moment had that mixture of joy and pain that jellyfish had so come to embody for me. As magical as it was that so many stars had aligned to allow me to experience it, and as magnificent as it felt to swim through the bloom, there was something wrong. The sheer mass, the bulk and breadth, of so many stinging jellyfish felt unhealthy, abnormal, even menacing. It felt like a party that was spinning out of control.

Bloom

After half a dozen years of chasing jellyfish around the world, I decided to do it one last time. But this time, it wouldn't be a sidebar to another family vacation or an excursion tagged on to one of Keith's business trips. I wouldn't cobble together frequent-flier miles or look for omens in fortune cookies. It would be a full-fledged jellyfish trip, and this time Keith was going to tag along with me.

It didn't hurt that this jellyfish trip was to Spain. I'd had my eye on a particular divot in the Mediterranean coast for some time. Called the Mar Menor, or the Little Sea, it is a triangular lagoon about the size of Washington D.C., that's sealed up with a strip of sand about fifteen miles long and a half mile wide. Before the 1960s, the region was dominated by sleepy fishing villages and scrappy farmers. But in the 1970s, two things happened to change the region and the jellyfish along with it.

First, tourists discovered the warm, calm waters. With a front porch on the Mediterranean and a backyard on the Mar Menor, developers realized that the strip of sandy land that sealed off the bay was high-euro property. They built so many high-rise hotels that, viewed from the shoreline today, it looks like a floating Las Vegas. During summer months, the human population in the region surges twentyfold with visitors swimming in the warm waters and strolling the white sand beaches. At the same time, industrial farmers discovered the region's three thousand–plus hours of sun a year. Large-scale irrigation projects and agricultural fields blossomed on the banks of the sea. Today, sewage from the tourists, brine from desalination plants, and fertilizers from the fields all wash into the bay. The runoff spikes the growth of the sea's microscopic plants, which feed the zooplankton, which feed the jellyfish. According to a report by the Regional Ministry of Agriculture, Water, and the Environment, the jellyfish in the Mar Menor are controlled by the amount of food available.

It's not that jellyfish were never seen in the Little Sea. Fishermen and locals remember seeing moon jellies every summer, and in much greater numbers than outside in the Mediterranean. But in the early 1990s, two new jellyfish appeared: the fried-egg jellyfish, *Cotylorhiza tuberculata*, and the barrel jellyfish, *Rhizostoma pulmo*. By 1993, the populations of these newcomers exploded, and the jellies have turned up every summer since. In 1996, there were 28 million jellies. By 2001, populations surged to 70 million and the regional government became alarmed. It established a large-scale program called Medusas to combat the jellyfish. While the Mar Menor's was one of the first, such jellyfish-control programs are increasing in number, with government agencies supporting mitigation efforts in many places, including Majorca, Malta, Australia, and New Jersey. The Mar Menor offered a glimpse into a future when jellyfish blooms are a chronic problem.

Keith and I checked into a simple hotel in the beach town of San Pedro del Pinatar, where we met up with Cristina Mena, who had administered the Medusas program when it was first established. Cristina gave off a playful energy. Her straight brown hair was pulled back in a loose ponytail. She had a wide face and crinkles near her eyes from laughing and sunshine. We strolled together along the more industrial section of the port, where working skiffs outnumbered elegant sailboats and where nets were strewn out, drying in the sun. We were looking for fishermen who could tell us about jellyfish in the lagoon. Suddenly, Cristina started waving broadly. Three men walked toward us, one younger and taller and two graying.

"We just got lucky," she said. I nodded because I had pretty much gotten used to the right people appearing in my search for jellyfish.

The younger man was Victor Moreno. Originally from the Canary Islands, he was one of the four Medusas divers who installed and maintained the jellyfish nets all summer. I asked him how the installation was going this year, and Victor told us that so far about half the nets had been put in the water. The rest—he pointed at the green mesh covering the dock—were piled around us. Victor offered to take us out on his boat to view the installed nets. But first, Cristina said—and Victor agreed—we needed a beer and some tapas.

We walked into a modest bar decorated only with a view of the sea and packed with locals. The tapas selection was arranged across the metal bar's countertop, and Cristina went to work ordering. Beer came first and then food arrived in a torrent. Mussels in olive oil, tiny salty fried squid, larger squid rings, octopus salad, delicate cod, briny fish eggs, a pretzel-like bread topped with potato salad and an anchovy. I felt déjà vu from Tsushima Island. Keith gave me a look that said, *This is incredible.* There was much too much to eat, and it was all delicious. Now he understood: This is what happens when you go in search of jellies.

After we were stuffed and sated, we boarded Victor's boat, which was well-worn but sturdy. A little bit of bird poop littered the deck, and some of the rails were weathered completely away, but the motor started with a smooth roar, and Victor smiled with pride. "The sound of 125 horses," he said.

We cruised across the green water toward the northern edge of the lagoon. There the nets extended from the shore outward about the length of two football fields. They were secured in the corners with yellow buoys and tied to concrete moorings on the seafloor. A string of lead weight held the net in place along the bottom. "We found that the jellyfish can get inside the net from underneath, so we have to make sure it touches the seafloor," Victor explained. As I'd learned in Japan, jellyfish know how to dive.

We puttered along nets that stretched the distance of more football fields, with narrow passages for boat ramps. All together, the nets cordoned off twenty-seven miles of shoreline and sixty beaches. With so much net, four divers were employed to constantly monitor their condition. Rips, which happen all the time, were hand sewn with a plastic needle and line. Besides the four divers, six boats, each with a captain and crew, trawled the surface for jellies every day. They offloaded their haul to a large barge, which dumped the jellyfish in a ditch near the airport, where they mostly evaporated. Two safety officers and one manager completed the Medusas team. It was a big operation with a big price tag—400,000 euros a year, according to local reports.

"Does it work?" I asked.

"Yes," Cristina answered. "The people who swim in the nets don't get stung. And this year there are very few jellyfish. For some reason there are huge numbers of shrimp instead. It's very odd."

I didn't find it that odd. I'd come to appreciate that the very unpredictability of jellyfish is all that's truly predictable.

We drifted along under the warm Spanish sun, feeling happy. My belly was full. I was relaxed by the beer. Keith was next to me. The visit had come together spontaneously and beautifully. We were laughing with people we'd just met and who generously shared food and fuel and facts with us.

But as always, jellyfish helped me to notice both the perfect and the imperfect. Watching people swim behind nets was wrong. They looked so caged in, so contained. It was as if the swimmers had removed themselves from the seas they had come to bathe in, boxed away from the freedom that the seas represent. I couldn't push away the feeling that if this was the future, we were creating our own prison.

I timed the trip to Spain strategically so that besides visiting the Mar Menor, I could also attend the Fifth International Jellyfish Blooms Symposium, which was held in Barcelona. Fifteen years earlier, Monty Graham had hosted the First International Jellyfish Symposium in Alabama, with just about fifty attendees, "and that was everyone in the world studying jellyfish," he told me. While everyone in the world who studied jellyfish wasn't going to be in Barcelona, certainly most of them were. Nearly everyone I'd met along the way was attending: Monty Graham, Jenny Purcell, and Keith Bayha, who studied jellyfish genetics; Wyatt Patry, who raised jellyfish for the Monterey Bay Aquarium; Stefano Piraino, who studied the immortal jellyfish; and Antonella Leone, who studied jellyfish as food. Lucas Brotz and Shin-ichi Uye, neither of whom I'd seen since Japan, were also coming. And I'd have the chance to meet Angel Yanagihara, who had protected Diana Nyad from box jellies. I could also reconnect with Amit Lotan, whose sting block had protected me from jellyfish stings in Israel. In a decade and a half, the number of attendees at the Jellyfish Blooms Symposium had more than quadrupled to include researchers from around the world, from every continent except Antarctica. It was the largest jellyfish gathering ever.

Through the years, I have assembled a by-no-means comprehensive list of the words for jellyfish in many different languages. My travels made collecting a few easy. In Spain, Italy, and France, a jellyfish is named for the misunderstood mythological Medusa. Similarly, Israel calls jellies *meduzot*. In Latin America, besides *medusas*, jellies are called *malaguas* or *agua mala*, meaning "bad water." Sometimes they are even called *lagrimas de mar*, which pulls at the heart with its meaning: "tears of the sea." More encouragingly, in Portuguese, jellyfish are known as *água-viva*, or "living water." I give the award for the most creative of the Romance-language names to *aguacuajada*, which means "curdled water" in Spanish. I'm stuck with the image of a flipped-over cottage cheese container pulsing away.

Jellyfish are called *kurage* in Japan, *sua* in Vietnamese. In China, "jellyfish" is *haizhe*. The word is written in two characters. The left part of the left character means "water," and the right side of that character means "mother." Together, the water-mother character represents the sea. The lower half of the right character means "bug." So jellyfish are known as sea bugs.

In Farsi, jellyfish are called *arood darya-i*, meaning "sea bride." I'm guessing that's for the long trailing oral arms that appear so veil-like. And in Turkish, they take on an even more powerful female role. Called *deniz anası*, jellyfish are likened to the mother of the sea. At the tip of India in Kerala, jellyfish are called *kadalchori*, which translates as "sea itch," which could be an understatement: There are known to be fierce stingers in those waters. In Indonesian, they are *ubur-ubur*.

In Norway, jellyfish are called *manet*, and there is more than one type. You don't want to meet up with *brennmanet*, because it will burn you. But you can pick up and play with the clear *glassmanet*, which look to be the common moon jellyfish. In Danish, a jellyfish is masculine: *vandmand*, or "water man." But my favorite of all is the name I wish I'd known when my kids were still small. It comes from the Welsh slang *psygod wibbly wobbly*, or "wiggle wiggle fish."

The names span our sensibility, from nurturing to evil and from mythical to banal. Unlike any other creature I know, jellyfish manage to waft back and forth across the line between the divine and the demonic. Despite all I've learned about jellyfish in the years that I've spent with them, these names are as good a summary as any of how I feel about these incredible creatures: Jellyfish are at the same time a reminder of our intimate connection to our oceans and the perils of our inaction; jellyfish are ultimately a symbol of potential.

As he had in Japan, Monty Graham opened the proceedings. Since I'd first met him, Monty had moved to more of an administrative position. He admitted that he didn't recognize many attendees and he'd had to catch up on much of the literature in preparation for his talk. That, he said, was a good thing, because it pointed to growth in the field. Monty's topic was the connection between jellyfish and human well-being. He began by introducing a social-science concept, Maslow's hierarchy of human needs, represented by a pyramid with the most basic human needs, such as food and safety, at the bottom, and more developed requirements, such as love and belonging, esteem, and self-actualization, at the top. Monty pointed out that jellyfish play roles at the basic levels, influencing food via fisheries and safety through jellyfish's contribution to functional ecosystems and the carbon cycle of our planet, not to mention their potential for harm from their stings. I would argue that they also play a role at the higher levels, unquestionably being a source of growth and strength for me and a muse for all who are captivated by their beauty and alien grace.

I ran into Jenny Purcell in the hallway in between sessions. I asked if she remembered me, and the conversation we'd had when she was in Qingdao, when she told me to write a book that mattered. She apologized and said she did not. I wasn't surprised. That her words had stuck with me for so many years, that they had been a driving force behind my work, shows how the little things we do can spin off and have unexpected consequences.

In her plenary talk the next day, Jenny emphasized the lessons she'd learned throughout her long career studying jellies. She explained that she'd always tried to find the most direct quantitative methods for understanding the creatures. But there's still so much we don't know, she said. Where are the polyps? How fast do jellies grow? When do they die? What triggers blooms? What role do predators play? How about parasites? She pointed out new technologies that give indirect measures of jellyfish, like acoustics, video, and aerial photography—not to mention the reports to jellyfish watch websites by you and me and anyone who goes to a beach or rides a boat on the sea. We, as a group, are known as citizen scientists, and sharing our information with these websites bulks up meager databases and is an invaluable tool for answering the critical questions about jellies on a global scale.

Jenny ended by mentioning the environmental classic by Rachel Carson, *Silent Spring*, which she said had influenced her as a child and probably pushed her to become a biologist. It was now the fiftieth anniversary of the book, which had sounded the alarm about the dangers of DDT and helped kick off the environmental movement in the 1960s, leading to the Clean Air and Clean Water acts that still protect us today. Yet the damages to our planet continue to accumulate, Jenny reminded the jellyfish scientists at the meeting. The human population continues to increase, compounding the environmental problems with overfishing, pollution, coastal development, increased shipping, and warming waters. All of these have the potential to benefit jellyfish.

At sixty-five but looking at least twenty years younger, Shin-ichi Uye had aged into mandatory retirement since the last time I saw him. He was no longer actively teaching at Hiroshima University, but his retirement had a silver lining, he told me. Now he was free to travel as much as he wanted, including to this meeting and several others in Europe. Ever the philosopher-scientist, Shin referred to President Obama's historic visit to Hiroshima four days earlier: "Obama has spoken of a nuclear-free future. A jellyfish-free future is unrealistic," he began. Jellyfish, of course, are no nuclear bomb, but our problems with jellyfish are certainly problems we have brought on ourselves. And similarly, only by working together do we have the power to fix them.

All of these eminent jellyfish scientists I had met through the years had taken the opportunity at this meeting, the largest meeting on jellyfish blooms ever held, to look back over their careers and assess what they had seen. Each in her or his own way had taken a step back from the everyday work of being a scientist in order to address the questions, problems, and issues surrounding not just jellyfish but our place here on Earth as well. Monty spoke of human needs and transcendence; Jenny appealed to needs of the planet; Shin emphasized our immense power. These words fell on the ears of hundreds of researchers from all corners of the world, who would go back home reinvigorated with the sense that to research jellyfish is not just to look at a creature unfamiliar and bizarre to most, but to study the planet and our place in it.

Toward the end of the meeting, one presenter talked of jellyfish science itself undergoing a bloom. There's much to back up the assertion. According to a 2012 study, the number of published scientific papers on jellyfish, salps, and comb jellies has doubled every decade since 1941. Today, hundreds of papers are published every year on these gelatinous creatures. As evidenced by the topics covered at the meeting, jellyfish research now extends well beyond ecology and fisheries, and includes genetics, toxicology, microbiology, taxonomy, numerical modeling, jellyfish husbandry, evolution, biogeochemistry, biomechanics, and aquaculture, among others. There was a talk on using jellyfish as fertilizer. There was a talk about the different microbial communities living on different populations of jellyfish, just as we harbor populations of bacteria in our guts. There was a talk about using jellyfish mucus to collect heavy metals from highly polluted water. There was an entire session on the benefits of jellyfish and another on developing treatments to stings. There were half a dozen talks on the untapped value of citizen science for reporting jellyfish sightings and gathering more data on their true abundances. A bloom is indeed happening, not just in growth of publications and in the fields of research into which jellyfish were spreading their tentacles, and not just in the growing numbers of jellyfish scientists and people interested in jellyfish, but also in the relevance of jellyfish to the world. Many, many questions remain, but jellyfish are finally starting to receive the attention they so deserve.

There's a fire in our bellies when we're young. When we take unconsidered risks. When we fall helplessly in love. When we dance on the beach until dawn. We throw ourselves headlong into the passion, zipping along the edges of the waves like the bioluminescence of a red tide. As we age, our fire continues to burn, though sometimes the live coals become buried under the ashes. The jellyfish helped me dig down to a fire inside.

While it's true that reaching the bottom of a wave can force us to change, that's not the only way change happens. I wasn't miserable when I found jellyfish. I was married to the best man I'd ever known, and our marriage was solid. I was blessed with two healthy, incredible children, who filled my world with delight. I had work. I had enough money. Things were fine. But a thousand little pricks to my psyche pushed me into the world of jellyfish. Those thousand little jabs, whether I realized they were there or not, that drove me toward this weird passion. None of them were crises—feeling bored at my job, missing the creativity of science, needing space from my kids, longing for meaning, needing to do something that mattered, recognizing that jellyfish are an important puzzle piece in our picture of the ocean, and a pressing sense that I wasn't raising my voice—and yet they were collectively significant. It turns out that those thousand little pinpricks can have the same impact as a deep dark moment. The jellyfish subdues its prey in the same way, with no dramatic statement, no overwhelming lunge. The jellyfish has no sharklike teeth or bearlike claws. A jellyfish is armed only with thousands of microscopic darts.

As humankind, we have emerged from the youth of civilization. We have struggled through our pubescence and reached the moment when our youthful good looks and passion are colliding with our need to grow up. Our collective emotions are still hot, however. Our dancing turns too simply to fighting, often for the wrong reasons. We fall in love easily, but often with things that don't matter or even harm us, things that numb us to the thousands of ever more discernible darts we are shooting at our own planet. This can lead to terrible mistakes, even self-destruction.

I realize now that I was wrong when I first swam in the coral reef and believed that biology was immutable, somehow not subject to human whim. We have reached a moment in history when we control the chemistry and biology of our planet. We are that powerful. But we are also endowed with gifts of even greater power. We have the capacity to communicate, to learn quickly, to change course, to create and to re-create, to make decisions for the health of the oceans, to speak up. We can protect this stunning planet we all share if we grow a collective spine. And we can. As a jellyfish scientist told me many years ago, we are an incredible species.

ACKNOWLEDGMENTS

I've often imagined a world where everyone gets to write down their acknowledgments—teachers, flight attendants, stockbrokers, tree trimmers, marine biologists, museum aquarists, jellyfishermen. I think the world would be a kinder and more humble place. I feel blessed with the privilege to say thank you to the people who have helped me.

My first thank yous go to the people who encouraged me when I didn't know how to start. Science writer Thomas Hayden was a long-distance mentor well before jellyfish swam into my life, and the entire science-writing community is enriched by his talent and integrity. Esther Mizrachi told me I was a writer before I was ready to hear it, and I am ever grateful for her years of enthusiastically rereading stories of jellies. Martin and Heather Kohout awarded me a writing residency at Madroño Ranch, where I wrote my first chapter. This book might not have begun without their belief in me. Julia Clarke joined me on that first trip and has been a constant sounding board about all things jellyfishy, scientific, and otherwise throughout the years. Margie Stohl somehow seemed to appear in my life every time I was in jellyfish crisis. Thank you for your constant encouragement and always sage advice. And thank you to the incomparable Dale Kiefer, for believing there was some ocean science inside me way back when.

Even though Austin is landlocked, it is an ecosystem schooling with groups that provide incredible support to writers. I have been fortunate to drift my way into several of them. To the Writing Group of the Shed—Esther Mizrachi, May Kuckro, Brittani Sonnenberg, Keb Frost, and Juliana O'Brien—thank you for your talent, energy, and love, which radiate into the universe and turn ideas into reality. To my Thursday Girls, Terry Benaryeh and Heila Lain, who know that looking good isn't nearly as important as sharing dreams. To Ladies, Lit, and Liquor, thank you for all the smart talk and laughs. And thanks to the Austin Science Communicators for keeping science hip and relevant in our hometown. I'd also like to acknowledge the staff and systems at the University of Texas at Austin Libraries and Susan Hovorka for making the world of jellyfish research so easily accessible. Thank you also to Frank and Jan King and to the Ferdman family for graciously providing informal but oh-so-necessary writing and research retreats through the years.

A number of people were of enormous help turning electronic words into an actual book. Alan Weisman sat with me on the front lawn of the Texas Capitol and listened to me gush about jellyfish. He then read the first draft of my book proposal and told me to keep going. Gioconda Parker provided me with a place to write that first proposal. Barbara Ras and Shannon Davies, both remarkable women overseeing fabulous academic presses, helped me believe that I could make this book real. Mollie Glick, my agent, is tougher and smarter than I could ever hope to be. She is a force I am so grateful to have on my side. And I give my gratitude to Courtney Young, my brilliant editor, who not only saw the potential in jellyfish but made this book stronger every time she touched its pages. Thank you also to the entire team at Riverhead for your enthusiasm about jellyfish.

In large part, this work is an ode to the people who spend their lives studying jellyfish, who decided to work in a dusty corner of marine biology, recognizing that without the corner, the house won't stand. These scientists shared hours of their time, responded enthusiastically to my out-of-the-blue e-mails, and talked endlessly about jellyfish with me. I'd like to thank them all for their work expanding our knowledge of our precious sea and its inhabitants. You met some of these experts the pages of this book, but others I wasn't able to include: Kelly Sutherland, University of Oregon; Laurence Madin and Richard Harbison, Woods Hole Oceanographic Institution; Richard Brodeur, Northwest Fisheries Science Center; Steven Haddock, Monterey Bay Research Institute; John Dabiri, Stanford University; Cathy Lucas, University of Southampton; Kylie Pitt, Griffith University; Anthony Moss, Auburn University; Tamara Shiganova, Russian Academy of Sciences; Thomas Holstein, Heidelberg University; Cornelia Jaspers, Technical University of Denmark; Martin Lilley,

Swansea University; Birgit Hoyer, Medizinisch-Theoretisches Zentrum; Zelda Montoya, Dallas Zoo; Alexander Semenov, Aquatilis; David Albert, independent scientist; Shachar Richter, Tel Aviv University; Chris Lynam, Centre for Environment, Fisheries and Aquaculture Science; Kara Dodge, University of New Hampshire; Ryosuke Makabe, Hiroshima University; Leonid Moroz, University of Florida; Amir Toren and Yossi Tal, Starlet Derma (now Monterey Bay Labs); Mary Beth Decker, Yale University; Julie Califf, Georgia Department of Natural Resources; Holly Binns, Pew Charitable Trusts; Jack Rudloe, Gulf Specimen Marine Laboratory; Eugene Raffield Jr., Raffield Fisheries; Ana Ramón Garcerán, University of Murcia; Anthony Amos, University of Texas Marine Sciences Institute; and Ann Molineux, University of Texas at Austin.

Science doesn't always drive emotion, but the fundamental goal of art is to make an emotional connection, and now more than ever, scientists need artists willing to tell their stories. I want to give a huge thanks to the artists who have helped to tell the stories of jellyfish in new and exciting ways: Marina Zurkow, Marcin Ignac, and Angela Haseltine Pozzi. And thank you to Stephanie Taiber, whose artistic perspective has inspired me through the years. I also extend enormous admiration to Rachel Ivanyi for turning my vision of jellyfish on the pages of this book into something glorious.

As I was finishing writing this book, fake facts became a fixture in our world, and I became even more keenly attuned to the need to tell the jellyfish's story as accurately as I could. The incisive fact-checker Jennifer Chaussee painstakingly worked through the book to ensure that the information is as correct as possible. I am also grateful to Shin-ichi Uye for being a science backstop and to Angel Yanigahara, Amit Lotan, Keith Bayha, and Lucas Brotz for reading versions of various chapters for accuracy. Some of the science in this book concerns ongoing research and debates. It has been my goal to relate leading lines of research and analysis that inform these critical discussions that make science so valuable. In the end, any remaining errors are my own.

Big thanks to the entire Stein clan: Jordan, Abbey, Max, Brad, and especially my sister Lynne, who made more than just the trips away from home possible. She was also my most patient and steadfast advocate. Thanks to Gail

Berwald for always telling me I could do anything and David Berwald for instilling in me a love of science and a passion for the natural world. And thanks to all the Berwalds and Ferns for cheering me on.

My greatest thanks go to the incredible people who live with me every day, who endured years of missed moments and distracted conversations because my mind had drifted off to the sea or my physical self had slipped away to visit labs or sit in front of the computer screen. Keith, you are my love; your patience, humor, and remarkable willingness to go on this jellyfish journey was the foundation for it all. You make me stronger than I ever imagined I could be. And to Ben and Isy, thank you for keeping me in the moment (as much as possible) and for being the best reasons for hope about our future that I know.

Notes

Epigraph

xi **"Among the most beautiful":** George J. Romanes, *Jelly-fish, Star-fish, and Sea-Urchins: Being a Research on Primitive Nervous Systems* (New York: D. Appleton, 1885), introduction, pp. 1–2, https://archive.org/stream/jellyfishstarfi04romagoog#page/n17/mode/2up.

1. If You Dare

6 **equations satellites use:** Juli Berwald et al., "Influences of Absorption and Scattering on Vertical Changes in the Average Cosine of the Underwater Light Field," *Limnology and Oceanography* 40 (1995): 1347–1357.

7 **according to Ovid:** "Perseus Tells the Story of Medusa," Ovid, *The Metamorphoses*, trans. Anthony S. Kline, book 4, lines 753–803, http://ovid.lib.virginia.edu/trans/Metamorph4.htm.

8 **close-up shots of marine organisms:** David Liittschwager, *A World in One Cubic Foot: Portraits of Biodiversity* (Chicago: University of Chicago Press, 2012).

8 **The piece David was working on:** Elizabeth Kolbert, "The Acid Sea," *National Geographic*, April 2011, http://ngm.nationalgeographic.com/print/2011/04/ocean-acidification/kolbert-text.

9 **increased the growth of the planted tube:** Jennifer E. Purcell, "Climate Effects on Formation of Jellyfish and Ctenophore Blooms: A Review," *Journal of the Marine Biological Association of the United Kingdom* 85 (2005): 461–76; Jennifer E. Purcell, "Environmental Effects on Asexual Reproduction Rates of the Scyphozoan *Aurelia labiata*," *Marine Ecology Progress Series* 348 (2007): 183–196.

9 **Beginning in 1963, off Namibia:** Christopher P. Lynam et al., "Jellyfish Overtake Fish in a Heavily Fished Ecosystem," *Current Biology*, 16 (2006): R492–R493.

9 **In the Black Sea:** Tunc Aybak, *Politics of the Black Sea: Dynamics of Cooperation and Conflict* (London: I. B. Taurus, 2001), pp. 136–137.

9 **In Japan . . . the giant jellyfish:** Masato Kawahara et al., "Unusual Population Explosion of the Giant Jellyfish *Nemopilema nomurai* (Scyphozoa: Rhizostomeae) in East Asian Waters," *Marine Ecology Progress Series* 307 (2006): 161–173.

9 **In the Philippines, jellyfish clogged:** "Jellyfish Blamed for Philippines Blackout," BBC World News Service, December 11, 1999, http://news.bbc.co.uk/2/hi/asia-pacific/560070.stm.

9 **The largest lagoon:** "Jellyfish Protection Nets Installed for Summer in the Mar Menor from 1st June," *Murcia Today*, May 28, 2015, http://murciatoday.com/jellyfish-protection-nets-installed-for-summer-in-the-mar-menor-from-1st-june_26065-a.html.

10 **Jellyfish Act of 1966:** H.R. 11475 (89th Congress, November 2, 1966): "An Act to Provide for the Control or Elimination of Jellyfish and Other Such Pests in the Coastal Waters of the United States," https://www.govtrack.us/congress/bills/89/hr11475/text.

10 **"Broad statements about global warming":** Steven H. D. Haddock, "Reconsidering Evidence for Potential Climate-Related Increases in Jellyfish," *Limnology and Oceanography* 53 (2008): 2759–2762.

12 **"a golden age of gelata":** Steven H. D. Haddock, "A Golden Age of Gelata: Past and Future Research on Planktonic Ctenophores and Cnidarians," *Hydrobiologia* 530 (2004): 549–556.

13 **blue-water diving:** Steven H. D. Haddock and John N. Heine, *Scientific Blue-Water Diving* (La Jolla, CA: Sea Grant College Program, 2005).

13 **creatures called hyperiid amphipods:** Nicholas E. C. Fleming et al., "Scyphozoan Jellyfish Provide Short-Term Reproductive Habitat for Hyperiid Amphipods in a Temperate Near-Shore Environment," *Marine Ecology Progress Series* 510 (2014): 229–240.

13 **baby lobsters:** Michiya Kamio et al., "Phyllosomas of Smooth Fan Lobsters (*Ibacus novemdentatus*) Encase Jellyfish Cnidae in Peritrophic Membranes in Their Feces," *Plankton & Benthos Research* 11 (2016): 100–104.

13 **diving birds:** N. N. Sato et al., "The Jellyfish Buffet: Jellyfish Enhance Seabird Foraging Opportunities by Concentrating Prey," *Biology Letters* 11 (2015).

13 **submersibles and remotely operated vehicles (ROVs):** Jennifer E. Purcell, "Extension of Methods for Jellyfish and Ctenophore Trophic Ecology to Large-Scale Research," *Hydrobiologia* 616 (2011): 23–50.

15 **meet virtually face-to-face:** I interviewed Jenny Purcell by Skype on April 5, 2012.

2. What's Your Agenda?

22 **conference on jellyfish blooms:** There have been five International Jellyfish Bloom Symposia. The first was held in Gulf Shores, Alabama, in 2000; the second, in Gold Coast, Queensland, Australia, in 2007; the third in Mar del Plata, Argentina, in 2010; the fourth in Hiroshima, Japan, in 2013; and the fifth in Barcelona, Spain, in 2016.

23 **He fired back an e-mail:** This e-mail was received on February 11, 2011.

23 **conversation started off awkwardly:** This conversation was recalled from written notes taken during the interview with Monty Graham on March 14, 2011, at Dauphin Island Sea Lab.

24 **"how jellyfish affect people":** Much of what Monty Graham told me was later published: William M. Graham et al., "Linking Human Well-being and Jellyfish: Ecosystem Services, Impacts, and Societal Responses," *Frontiers in Ecology and the Environment* 12 (2014): 515–523.

24 **July 2015 in Israel:** Ruth Schuster, "Israeli Power Plant Fights Off Giant Jellyfish Swarm," *Haaretz*, June 25, 2015, http://www.haaretz.com/israel-news/.premium-1.662930.

24 **October 2013 in Sweden:** Associated Press, "Jellyfish Clog Pipes of Swedish Nuclear Reactor Forcing Plant Shutdown," *The Guardian*, October 1, 2013, https://www.theguardian.com/world/2013/oct/01/jellyfish-clog-swedish-nuclear-reactor-shutdown.

24 **2012 near Santa Barbara:** "Diablo Canyon Nuclear Plant in California Knocked Offline by Jellyfish-like Creature Called Salp," NBC News, April 27, 2012, http://usnews.nbcnews.com/_news/2012/04/27/11432974-diablo-canyon-nuclear-plant-in-california-knocked-offline-by-jellyfish-like-creature-called-salp.

24 **2011 in Scotland, Israel, and Japan:** Basasubramanyam Seshan,"Jellyfish Invade Four Nuclear Reactors in Japan, Israel, Scotland," *International Business Times*, July 10, 2011, http://www.ibtimes.com/jellyfish-invade-four-nuclear-reactors-japan-israel-scotland-photos-707777.

24 **power outages . . . in the Philippines:** "Jellyfish Blamed for Philippines Blackout," BBC News, December 11, 1999, http://news.bbc.co.uk/2/hi/asia-pacific/560070.stm.

24 **a senior scientist:** The quotation is from jellyfish scientist Shin-ichi Uye, whom I would meet during my trip to Japan in 2012.

24 **most embarrassing jellyfish clog:** Drew Creighton, "Early Jellyfish Bloom in Moreton Bay," *Brisbane Times*, November 22, 2016, http://www.brisbanetimes.com.au/queensland/early-jellyfish-bloom-in-moreton-bay-20161122-gsv8bx.html.

25 **in Oman and Israel:** Sunil K. Vaidya, "Jellyfish Choke Oman Desalination Plants," *Gulf News*, May 6, 2003, http://gulfnews.com/news/uae/general/jellyfish-choke-oman-desalination-plants-1.355525; Bella S. Galil, Nurit Kress, and Tamara A. Shiganova, "First Record of *Mnemiopsis leidyi* A. Agassiz, 1865 (Ctenophora; Lobata; Mnemiidae) off the Mediterranean Coast of Israel," *Aquatic Invasions* 4 (2009): 357–360.

26 **The deadliest jellyfish:** Lisa-ann Gershwin, "Box Jellyfish & Irukandji Deaths in Australia," *Australian Marine Stinger Advisory Services*, http://www.stingeradvisor.com/boxydeaths.htm; P. J. Fenner, "*Chironex fleckeri*—the North Australian Box-Jellyfish," Marine-Medic, http://www.marine-medic.com.au/pages/biology/biologyBreakup/jellyfishChironex.pdf.

26 **A report from Scotland:** "Jellyfish as a Nuisance Species to Aquaculture," Scottish Government, http://www.gov.scot/Topics/marine/marine-environment/species/plankton/nuisance.

27 **successful invasion by a small, gelatinous animal:** General Fisheries Commission for the Mediterranean, "First Meeting of the GFCM Working Group on the Black Sea," January 2012.

27 **a ship leaving the Gulf of Mexico:** Sara Ghabooli et al., "Multiple Introductions and Invasion Pathways for the Invasive Ctenophore *Mnemiopsis leidyi* in Eurasia," *Biological Invasions* 13 (2011): 679–690.

28 **half a million tons:** Temel Oguz, Bettina Fach, and Baris Salihoglu, "Invasion Dynamics of the Alien Ctenophore *Mnemiopsis leidyi* and Its Impact on Anchovy Collapse in the Black Sea," *Journal of Plankton Research* 30 (2008): 1385–1397.

28 **total landings of fish in the Black Sea:** G. M. Daskalov et al., "Trophic Cascades Triggered by Overfishing Reveal Possible Mechanisms of Ecosystem Regime Shifts," *Proceedings of the National Academy of Sciences of the United States of America* 104 (2007): 10518–10523.

29 **$1 billion or more:** Tunç Aybak, ed., *Politics of the Black Sea: Dynamics of Cooperation and Conflict* (London: I. B. Tauris, 2001), pp. 136–137.

29 ***Beroe:*** This blog post, besides being a delight to read, has astonishing video of *Beroe* engulfing *Mnemiopsis*: Rebecca Helm, "The Rainbow-Covered Animal with a Dark Side," *Deep Sea News*, November 5, 2014, http://www.deepseanews.com/2014/11/the-rainbow-covered-animal-with-a-bite.

30 ***Beroe ovata* appeared:** Tamara A. Shiganova et al., "The New Invader *Beroe ovata* Mayer 1912 and Its Effect on the Ecosystem in the Northeastern Black Sea," *Hydrobiologia* 451 (2001): 187–197.

30 **invasive nature of *Mnemiopsis*:** J. H. Costello et al., "Transitions of *Mnemiopsis leidyi* (Ctenophora: Lobata) from a Native to an Exotic Species: A Review," *Hydrobiologia* 690 (2012): 21–46.

31 **jellyfish in Namibian waters:** Christopher P. Lynam et al., "Jellyfish Overtake Fish in a Heavily Fished Ecosystem," *Current Biology* 16 (2006): R492–R493.

32 **overfishing was clearly a culprit:** Jean-Paul Roux et al., "Jellification of Marine Ecosystems as a Likely Consequence of Overfishing Small Pelagic Fishes: Lessons from the Benguela," *Bulletin of Marine Science* 89 (2013): 249–284.

33 **"Jellyfish play potentially major":** Christopher P. Lynam et al., "Jellyfish Overtake Fish in a Heavily Fished Ecosystem," *Current Biology*, 16 (2006): R493.

33 **jellyfish-based problems:** Jennifer E. Purcell, "Jellyfish and Ctenophore Blooms Coincide with Human Proliferations and Environmental Perturbations," *Annual Review of Marine Science* 4 (2012): 209–235.

34 **Jellyfish and human perception:** Cathy H. Lucas et al., "Gelatinous Zooplankton Biomass in the Global Oceans: Geographic Variation and Environmental Drivers," *Global Ecology and Biogeography* 23 (2014): 701–714.

38 **humble jellyfish might act:** Brad J. Gemmell et al., "Can Gelatinous Zooplankton Influence the Fate of Crude Oil in Marine Environments?" *Marine Pollution Bulletin* 113 (2016): 483–487.

3. Jellyfish Salad

41 **jellyfish, which has been eaten in Asia:** Y.-H. Peggy Hsieh, Fui-Ming Leong, and Jack Rudloe, "Jellyfish as Food," *Hydrobiologia* 451 (2001): 11–17.

43 **"Of sea-nettles there are two":** Aristotle, *The History of Animals*, trans. D'Arcy Wentworth Thompson, book 4, part 6, http://classics.mit.edu/Aristotle/history_anim.4.iv.html.

43 **"This medusa is abundant in the Inland Sea of Japan":** Alfred Goldsborough Mayer, *Medusae of the World* (Washington, DC: Carnegie Institution of Washington, 1910), vol. 3, p. 706, https://archive.org/stream/medusaeofworld03mayo#page/706/mode/2up.

44 **"*Rhopilema hispidum* is on sale":** John A. Williamson et al., eds., *Venomous and Poisonous Marine Animals: A Medical and Biological Handbook* (Sydney: University of South Wales Press, 1996), p. 212.

44 **I caught up with Wynn:** I interviewed Wynn Gale by phone on April 4, 2016.

45 **A study of Georgia jellyballers:** James W. Page, "Characterization of Bycatch in the Cannonball Jellyfish Fishery in the Coastal Waters off Georgia," *Marine Coastal Fisheries* 7 (2015): 190–199.

46 **Southeast's jellyfishing industry:** Justin Nobel, "Jellyfish: It's What's for Dinner," *Modern Farmer*, September 15, 2014, http://modernfarmer.com/2014/09/jellyfish-whats-dinner; Mary Landers, "Coastal Georgia Shrimpers Turn to Jellyfish to Make Money," *Savannah Morning News*, posted May 15, 2011, updated April 20, 2016, http://savannahnow.com/news/2011-05-15/coastal-georgia-shrimpers-turn-jellyfish-make-money#.

46 **managed by April Harper:** I interviewed April Harper by phone on April 1, 2016.

47 **people were eating unsafe levels of alum:** Yuan Duanduan and Li Yifan, "A Third of China's Population Eating Too Much Aluminium," *China Dialogue*, originally published in *Southern Weekend*, July 23, 2014, https://www.chinadialogue.net/article/show/single/en/7158-A-third-of-China-s-population-eating-too-much-aluminium.

48 **A blog called *Deep End Dining*:** "All Those Jellyfish. So Little Time. PB&JF." *Deep End Dining*, March 4, 2007, http://www.deependdining.com/2007/03/all-those-jellyfish-so-little-time-pb.html.

49 **A serving of jellyfish contains:** Nutrition information for cannonball jellyfish is from Golden Island, http://www.giiseafood.com/ct-menu-item-7.

50 **two companies in Europe have begun harvesting jellyfish:** Jellagen is based in the UK, and Javenech is based in France.

50 **In studies done in China:** Jian Fan, Zhuang Yongliang, and Bafang Li, "Effects of Collagen and Collagen Hydrolysate from Jellyfish Umbrella on Histological and Immunity Changes of Mice Photoaging," *Nutrients* 5 (2013): 223–233.

50 **analyzed three species of jellyfish:** Antonella Leone et al., "The Bright Side of Gelatinous Blooms: Nutraceutical Value and Antioxidant Properties of Three Mediterranean Jellyfish (Scyphozoa)," *Marine Drugs* 13 (2015): 4654–4681.

50 **When I spoke to her on the phone:** I spoke with Antonella Leone on April 6, 2016.

4. Missing Polyp

53 **The first meeting on my schedule was with Alex Andon:** My interview with Alex Andon was on May 25, 2011, near San Francisco.

55 **a little jellyfish sex ed:** For an excellent and comprehensive treatise on the biology of jellyfish, see Mary Arai, *A Functional Biology of Scyphozoa* (London: Chapman & Hall, 1997).

57 **a documentary called *Rise of the Jellyfish*:** Carsten Oblaender, producer/director, *Rise of the Jellyfish* (Story House Productions, for Discovery Channel, 2011).

58 **the very spinelessness of the jellyfish:** Arai, *A Functional Biology of Scyphozoa*, pp. 16–19, 133–136.

59 **insensitive to oxygen concentration:** Michael Grove and Denise L. Breitburg, "Growth and Reproduction of Gelatinous Zooplankton Exposed to Low Dissolved Oxygen," *Marine Ecology Progress Series* 301 (2005): 185–198.

59 **growing just as well in water with low oxygen:** Mary-Elizabeth C. Miller and William M. Graham, "Environmental Evidence That Seasonal Hypoxia Enhances Survival and Success of Jellyfish Polyps in the Northern Gulf of Mexico," *Journal of Experimental Marine Biology and Ecology* 432–433 (2012): 113–120.

60 **"A white belt in Kuragedo":** Chad Widmer, "On the Art of Kuragedo," *International Jellyfish Conference*, December 2, 2016, http://international-jellyfish.net/2016/02/12/on-the-art-of-kuragedo. See also Chad's outstanding book on rearing jellyfish, *How to Keep Jellyfish in Aquariums: An Introductory Guide for Maintaining Healthy Jellies* (Tucson: Wheatmark, 2008).

63 **I first found Lucas Brotz:** My interview with Lucas Brotz was conducted on April 27, 2012, by phone.

63 **a controversial paper:** Lucas Brotz et al., "Increasing Jellyfish Populations: Trends in Large Marine Ecosystems," *Hydrobiologia* 690 (2012): 3–20.

63 **Coastal development is rampant:** Carlos M. Duarte et al., "Is Global Ocean Sprawl a Cause of Jellyfish Blooms?," *Frontiers in Ecology and the Environment* 11 (2012): 91–97.

63 **studies of wild moon jelly polyps in Puget Sound:** Jennifer E. Purcell, Richard A. Hoover, and Nathan T. Schwarck, "Interannual Variation of Strobilation by the Scyphozoan *Aurelia labiata* in Relation to Polyp Density, Temperature, Salinity, and Light Conditions *in Situ*," *Marine Ecology Progress Series* 375 (2009): 139–149.

64 **locations of the polyps that produce those medusae remain mysterious:** Cathy H. Lucas, William M. Graham, and Chad Widmer, "Jellyfish Life Histories: Role of Polyps in Forming and Maintaining Scyphomedusa Populations," *Advances in Marine Biology* 63 (2012): 133–196.

65 **Scientists in the Netherlands recently used DNA markers:** Lodewijk van Walraven et al., "Where Are the Polyps? Molecular Identification, Distribution and Population Differentiation of *Aurelia aurita* Jellyfish Polyps in the Southern North Sea Area," *Marine Biology* 163 (2016): 172.

5. In Jelly Genes

67 **James "Whitey" Hagadorn told me:** This interview took place at the Denver Museum of Nature & Science on August 5, 2011.

68 **medusae fossils:** Graham A. Young and James W. Hagadorn, "The Fossil Record of Cnidarian Medusae," *Palaeoworld* 19 (2010): 212–221.

68 **a 560-million-year-old fossil:** Alexander G. Liu et al., "*Haootia quadriformis* n. gen., n. sp., Interpreted as a Muscular Cnidarian Impression from the Late Ediacaran Period (Approx. 560 Ma)," *Proceedings of the Royal Society B* 281 (2014): doi: 10.1098/rspb.2014.1202.

68 **small group of jellies called staurozoa:** Claudia Mills at the University of Washington has been assembling the most up-to-date information about staurozoa, https://faculty.washington.edu/cemills/Stauromedusae.html.

70 **I had sat down with Keith Bayha:** Quotations in this section are from an interview with Keith Bayha at Dauphin Island Sea Lab on March 14, 2011.

71 **His stunning and ornate drawings of plants and animals:** It's really worth looking at Haeckel's stunning drawings of jellyfish and other organisms, which are found in his classic art books *Kunstformen der Natur* and *Kunstformen aus dem Meer*, both reissued in a single volume (Munich: Prestel Verlag, 2012).

72 **"I had this strange":** Ernst Haeckel, *System der Acraspeden*, part 2 of *Das System der Medusen* (Jena: Gustav Fischer, 1880), p. 633, translation provided by Keith Bayha.

73 **A description of Stiasny characterizes him:** L. B. Holthius, *Rijksmuseum van Natuurlijke Historie 1820–1958*, pp. 119, 131.

73 **"In 1940, when the German heel":** Wim Vervoort, "Gustav Albert Stiasny, December 10, 1877–June 12, 1946," *Zoologische Mededelingen*, 30 (1950): 257–267, http://www.repository.naturalis.nl/document/149369.

74 **"For many years":** Gustav Stiasny, "Uber *Drymonema dalmatina* Haeckel," *Zoologische Jahrbücher Abteilung für Anatomie und Ontogenie der Tiere* 66 (1940): 437–462, translation provided by Keith Bayha.

75 **"Let us recall":** P. F. S. Cornelius, "The Hydrozoan Work of Prof. Wim Vervoort," *Zoologische verhandelingen Leiden* 323 (1998): 17–19, http://repository.naturalis.nl/document/148929.

75 **"The occurrence of *Drymonema dalmatina*":** Stiasny, "Uber *Drymonema dalmatina* Haeckel."

75 **Larson noticed one rosy blob:** Ronald J. Larson, "First Report of the Little-Known Scyphomedusa *Drymonema dalmatinum* in the Caribbean Sea, with Notes on Its Biology," *Bulletin of Marine Science* 40 (1987): 437–441.

77 **For this analysis, fifty-five:** Keith M. Bayha et al., "Evolutionary Relationships Among Schyphozoan Jellyfish Families Based on Complete Taxon Sampling and Phylogenetic Analyses of 18S and 28S Ribosomal DNA," *Integrative and Comparative Biology* 50 (2010): 436–455.

77 **the same species appeared:** Keith M. Bayha et al., "Predation Potential of the Jellyfish *Drymonema larsoni* Bayha & Dawson (Scyphozoa: Drymonematidae) on the Moon Jellyfish *Aurelia* sp. in the Northern Gulf of Mexico," *Hydrobiologia* 690 (2012): 189–197.

78 **the jellyfish became the definition:** Keith M. Bayha and Michael N. Dawson, "New Family of Allomorphic Jellyfishes, Drymonematidae (Scyphozoa, Discomedusae), Emphasizes Evolution in the Functional Morphology and Trophic Ecology of Gelatinous Zooplankton," *Biological Bulletin* 219 (2010): 249–267.

79 **Defining the moon jellies:** Michael N. Dawson, "Macro-Morphological Variation Among Cryptic Species of the Moon Jellyfish, *Aurelia* (Cnidaria: Scyphozoa)," *Marine Biology* 143 (2003): 369–379; Simonetta Scorrano et al., "Unmasking *Aurelia* Species in the Mediterranean Sea: An Integrative Morphometric and Molecular Approach," *Zoological Journal of the Linnean Society*, 2016, doi: 10.1111/zoj.12494.

79 **the number of jellyfish species is probably double:** Ward Appeltans et al., including Michael N. Dawson, "The Magnitude of Global Marine Species Diversity," *Current Biology* 22 (2012): 2189–2202.

79 **"Cryptic taxa . . . likely result":** Bayha and Dawson, "New Family of Allomorphic Jellyfishes, Drymonetidae (Scyphozoa, Discomedusae), Emphasizes Evolution," 264.

80 **How can we predict:** In August 2014, for the first time in seventy years, *Drymonema* appeared in

the Adriatic Sea, which raises the question, What has it been doing all that time? Nick Squires, "Giant Jellyfish Spotted in the Adriatic for First Time Since Second World War," *The Telegraph*, August 7, 2014, http://www.telegraph.co.uk/news/worldnews/europe/italy/11019006/Giant-jellyfish-spotted-in-the-Adriatic-for-first-time-since-Second-World-War.html.

80 **But in 2008, scientists delving into the origins of animals:** Casey W. Dunn et al., "Broad Phylogenomic Sampling Improves Resolution of the Animal Tree of Life," *Nature* 452 (2008): 745–749.

81 **The results agreed with the first study:** Joseph F. Ryan et al., "The Genome of the Ctenophore *Mnemiopsis leidyi* and Its Implications for Cell Type Evolution," *Science* 342 (2013): 1336.

81 **The problems, the new analyses explained, stemmed from comb jelly genes:** Davide Pisani et al., "Genomic Data Do Not Support Comb Jellies as the Sister Group to All Other Animals," *Proceedings of the National Academy of Sciences of the United States of America* 112 (2015): 15402–15407.

81 **complexity isn't such a simple thing:** Nathan V. Whelan et al., "Error, Signal, and the Placement of Ctenophora Sister to All Other Animals," *Proceedings of the National Academy of Sciences of the United States of America* 112 (2015): 5773–5778; Casey W. Dunn, Sally P. Leys, and Steven H. D. Haddock, "The Hidden Biology of Sponges and Ctenophores," *Trends in Ecology & Evolution* 30 (2015): 282–291.

6. Robojelly

85 **The lecturer mentioned a book called *Life in Moving Fluids:*** Steven Vogel, *Life in Moving Fluids: The Physical Biology of Flow* (Princeton, NJ: Princeton University Press, 1981).

87 **It's a real joy to sit and talk to these scientists:** I spoke with Sean Colin and Jack Costello in their lab in Woods Hole on July 26, 2011, June 28, 2013, and August 14, 2015.

88 **silicone flaps:** Sean P. Colin et al., "Biomimetic and Live Medusae Reveal the Mechanistic Advantages of a Flexible Bell Margin," *PLoS One* 7 (2012): e48909.

89 **studied videos of fifty-nine different animals that swim or fly:** Kelsey N. Lucas et al., "Bending Rules for Animal Propulsion," *Nature Communications* 5 (2014): 3293.

90 **Jellyfish are the most efficient swimmers:** Brad J. Gemmell et al., "Passive Energy Recapture in Jellyfish Contributes to Propulsive Advantage over Other Metazoans," *Proceedings of the National Academy of Sciences of the United States of America* 110 (2013): 17904–17909.

90 **a life-size droid called Cyro:** Alex A Villanueva et al., "Biomimetic Autonomous Robot Inspired by the *Cyanea capillata* (Cyro)," *Bioinspiration & Biomimetics* 8 (2013): 046005.

91 **The creation, called a medusoid:** Janna C. Nawroth et al., "A Tissue-Engineered Jellyfish with Biomimetic Propulsion," *Nature Biotechnology* 30 (2012): 792–797.

92 **region of low pressure:** Brad J. Gemmell et al., "Suction-Based Propulsion as a Basis for Efficient Animal Swimming," *Nature Communications* 6 (2015): 8790.

7. Seeing What's Not There

96 **a jellyfish scientist named Richard Harbison:** I interviewed Richard Harbison on July 25, 2011, in his office in Woods Hole.

97 **Sönke told me that the best answer:** I interviewed Sönke Johnsen by phone on August 25, 2012.

98 **One way animals can maximize transparency:** Sönke Johnsen and Edith A. Widder, "The Physical Basis of Transparency in Biological Tissue: Ultrastructure and the Minimization of Light Scattering," *Journal of Theoretical Biology* 199 (1999): 181–198. Sönke has written a wonderful summary of the interactions between light and marine organisms in his approachable book *The Optics of Life: A Biologist's Guide to Light in Nature* (Princeton, NJ: Princeton University Press, 2012).

98 **Jellyfish are physically fat and optically skinny:** Sönke Johnsen, "Hidden in Plain Sight: The Ecology and Physiology of Organismal Transparency," *Biological Bulletin* 201 (2001): 301–318.

100 **"red is the new black":** Sönke Johnsen, "The Red and the Black: Bioluminescence and the Color of Animals in the Deep Sea," *Integrative Comparative Biology* 45 (2005): 234–246.

101 **red tide:** Raphael M. Kudela and William P. Cochlan, "Nitrogen and Carbon Uptake Kinetics and the Influence of Irradiance for a Red Tide Bloom off Southern California," *Aquatic Microbial Ecology* 21 (2000): 31–47.

102 **single-person submersible:** Sneed B. Collard III, *In the Deep Sea* (New York: Marshall Cavendish Benchmark, 2006).

103 **Edie has become an interpreter:** I spoke with Edie Widder by phone on January 14, 2016.

103 **There's a sea cucumber . . . There's a copepod:** Edith A. Widder, "Bioluminescence in the Ocean: Origins of Biological, Chemical, and Ecological Diversity," *Science* 328 (2010): 704–708. Edie has also given excellent TED videos on deep-sea bioluminescence: "Glowing Life in an Underwater World," April 2010, and "The Weird, Wonderful World of Bioluminescence," March 2011.

105 **a call for help like none other:** P. J. Herring and Edith A. Widder, "Bioluminescence of Deep-Sea Coronate Medusae (Cnidaria: Scyphozoa)," *Marine Biology* 146 (2004): 39–51.

107 **We need a NASA for the sea:** Edith Widder, "How We Found the Giant Squid," TED Talk, February 2013, https://www.ted.com/talks/edith_widder_how_we_found_the_giant_squid.

107 **Ocean Research & Conservation Association, or ORCA:** http://www.teamorca.org/orca/index.cfm.

107 **"more than 90 percent":** Edith Widder, "The Weird, Wonderful World of Bioluminescence," TED Talk, May 2011, https://www.ted.com/talks/edith_widder_the_weird_and_wonderful_world_of_bioluminescence/transcript?language=en.

8. Day-glo Jellies

109 ***"pulmone marino si confricetur lignum":*** The translation of the sentence is "If wood is thoroughly rubbed with *Pulmo marinus*, it seems to be on fire, so much so that a walking-stick, so treated, throws a light forward." *Pulmo marinus* is literally "sea lung," but it's possible this was the common Mediterranean jellyfish *Pelagia noctiluca*, which is known to glow brightly. Pliny the Elder, *Natural History*, vol. 8, books 28–32, trans. W. H. S. Jones (Cambridge, MA: Harvard University Press, Loeb Classical Library, 1963), p. 551, https://www.loebclassics.com/view/pliny_elder-natural_history/1938/pb_LCL418.551.xml.

109 **"and most of them shine":** Carl Linnaeus, *A General System of Nature, Through the Three Grand Kingdoms . . .* (London: Lackington, Allen, 1806), vol. 4, p. 121.

109 **"No; this was not the calm irradiation":** Jules Verne, *Twenty Thousand Leagues Under the Sea* (New York: Parents Magazine's Cultural Institute, 1964), pp. 171–172.

110 **In 1945, sixteen-year-old Osamu Shimomura:** Much more about the discovery of the chemistry of bioluminescence, including Shimomura's story, can be found in Vincent Pieribone and David F. Gruber, *Aglow in the Dark: The Revolutionary Science of Bioluminescence* (Cambridge, MA: Belknap/Harvard University Press, 2005).

114 **"The protein is like a gun":** Ibid., p. 70.

114 **these ancient animals hold the secret:** Principal investigator Kenneth C. Lerner et al., "Madison Memory Study: A Randomized Double-Blinded Placebo-Controlled Trial of Apoaequorin in Community-Dwelling Older Adults," *Quincy Bioscience*, August 1, 2016, http://www.prevagen.com/research. Note that this research is published by the product's sponsor, Prevagen, and not in a peer-reviewed journal. In early 2017, the Federal Trade Commission sued Prevagen for misleading the public. Prevagen denied the charges. See https://www.ftc.gov/news-events/press-releases/2017/01/ftc-new-york-state-charge-marketers-prevagen-making-deceptive.

116 **the worm's body glowed:** Martin Chalfie et al., "Green Fluorescent Protein as a Marker for Gene Expression," *Science* 263 (1994): 802–805, http://science.sciencemag.org/content/sci/263/5148/local/front-matter.pdf.

117 **This palette of fluorescence:** Roger Y. Tsein, "Constructing and Exploiting the Fluorescent Protein Paintbox," Nobel Lecture, December 8, 2008.

117 **The uses for the fluorescent proteins:** Marc Zimmer, "Fluorescent Proteins Light Up Science by Making the Invisible Visible," *The Conversation*, April 7, 2015, http://theconversation.com/fluorescent-proteins-light-up-science-by-making-the-invisible-visible-39272.

119 **"When scientists develop methods":** 2008 Nobel Prize in Chemistry, Popular Information, Nobelprize.org, http://www.nobelprize.org/nobel_prizes/chemistry/laureates/2008/popular.html. Oshamu Shimomura's lecture, given on December 8, 2008, can be viewed at https://www.nobelprize.org/mediaplayer/index.php?id=1066.

9. Jellyfish Sense

122 **two types of eyes:** Mary Arai, *A Functional Biology of Scyphozoa* (London: Chapman & Hall, 1997), p. 31.

123 **Changing light also makes jellies' eggs mature:** Vicki J. Martin, "Photoreceptors of Cnidarians: A Review," *Canadian Journal of Zoology* 80 (2002): 1703–1722.

124 **"You have to kiss them":** Quotations of Anders Garm in this section are from a phone conversation with him on September 24, 2014.

124 **"a bizarre cluster of different eyes" . . . "For such a minute eye":** Dan-Eric Nilsson et al., "Advanced Optics in a Jellyfish Eye," *Nature* 435 (2005): 7039.

125 **"The sharp focus of the lens". . . Anders used mazes:** Anders Garm et al., "Pattern- and Contrast-Dependent Visual Response in the Box Jellyfish *Tripedalia cystophora*," *Journal of Experimental Biology* 216 (2013): 4520–4529; Anders Garm, M. O'Connor, L. Parkefelt, and Dan-E Nilsson, "Visually Guided Obstacle Avoidance in the Box Jellyfish *Tripedalia cystophora* and *Chiropsella bronzie*," *Journal of Experimental Biology* 210 (2007): 3616–3623.

125 **landmarks for navigation:** Anders Garm, Magnus Oskarsson, and Dan-Eric Nilsson, "Box Jellyfish Use Terrestrial Visual Cues for Navigation," *Current Biology* 21 (2011): 798–803.

127 **the most important sense: proprioception:** Robert W. Baloh, *Vertigo: Five Physician Scientists and the Quest for a Cure* (Oxford: Oxford University Press, 2016).

128 **Together, the touch plate and the statocyst form:** Arai, *A Functional Biology of Scyphozoa*, p. 32.

128 **first . . . jellies in space:** This educational video (with classic 1990s stylings) produced by NASA about the mission provides footage of ephyrae in space: "From Undersea to Outer Space: The STS-40 Jellyfish Experiment," published on YouTube, October 11, 2012, https://www.youtube.com/watch?v=FvjVKAAvIn8.

129 **the spaceship *Columbia*:** I discussed the jellyfish mission with two astronauts by phone: Tamara Jernigan on June 22, 2015, and Millie Hughes-Fulford on September 3, 2015.

130 **chronic vertigo:** Dorothy B. Spangenberg et al., "Development Studies of *Aurelia* (Jellyfish) Ephyrae Which Developed During the SLS-1 Mission," *Advances in Space Research* 14 (1994): 239–247; Dorothy B. Spangenberg et al., "Graviceptor Development in Jellyfish Ephyrae in Space and on Earth," *Advances in Space Research* 14 (1994): 317–325.

130 **thyroxine might play a role in the development of statocysts:** Dorothy B. Spangenberg, "Role of Thyroxine in Space-Developed Jellyfish," *Final Report to the National Aeronautics and Space Administration*, grant no. NAG10-0178, September 18, 1995–September 17, 1997.

10. The Nerve of the Jellyfish

132 **When he was just two years old:** For Romanes's background, see Mabel Ringereide, "Romanes—Father and Son" *The Bulletin* 28 (1979): 35–46, http://post.queensu.ca/~forsdyke/romanes.htm#Genealogy.

132 **"You have done too splendid work":** Ethel Duncan Romanes, *The Life and Letters of George John Romanes* (London: Longmans, Green, 1896), p. 103.

133 **"crouching attitude" . . . "It is interesting to observe":** George's letter to Darwin about the experiment is on page 24 of *The Life and Letters of George John Romanes.* He wrote more extensively about it in George John Romanes, *Jelly-fish, Star-fish and Sea-Urchins: Being a Research on Primitive Nervous Systems* (New York: D. Appleton, 1885), pp. 108–111, from which the quotations in this paragraph are taken. See https://archive.org/stream/jellyfishstarfi04romagoog#page/n9/mode/2up.

133 **But how does the jellyfish "know":** George eventually understood that the animal has strands of nerves that run from the center of the bell down to the edges, like the seams in a skirt. If a prick happens on the edge of the bell, the nerve band closest to the prick sends a signal that it's been touched, and the cone-mouth can follow that signal.

134 **"I feel that it is desirable to touch":** Romanes, *Jelly-fish, Star-fish and Sea-Urchins*, p. 6.

134 **"fundamental experiments":** Ibid., pp. 26–36.

135 **wave of contraction caused by a jellyfish pacemaker:** For a comparison of Romanes's experiments and modern scientific understanding of jellyfish nervous systems, see Richard A. Satterlie, "Neuronal Control of Swimming in a Jellyfish: A Comparative Story," *Canadian Journal of Zoology* 80 (2002): 1654–1669.

138 **"the contraction waves starting from the ganglion":** Romanes, *Jelly-fish, Star-fish and Sea-Urchins*, p. 68.

138 **"Now in this experiment"**: Ibid., p. 71.

139 **"If the reader will imagine"**: Ibid.

139 **"the physiological continuity is maintained"**: Ibid., p. 80.

140 **"We probably have a physical explanation"**: Ibid., p. 87.

140 **Today's understanding of how we create memories:** For an accessible explanation of the way we build memories using nerves, see *Memory and Learning* by Paul Heideman, which he has made available for download: http://pdheid.blogs.wm.edu/44-2. (Full disclosure: I helped edit this text.)

143 **"I think it is desirable to append a list"**: Romanes, *The Life and Letters of George John Romanes*, p. 105. (Emphasis in original.)

143 **Darwin wrote back:** David Galton, "Did Darwin Read Mendel?" *Quarterly Journal of Medicine* 102 (2009): 587–589.

145 **"But the more I think about the whole thing"**: Romanes, *The Life and Letters of George John Romanes*, pp. 90–91.

11. Life's Limits

150 **There I would meet Stefano Piraino and Ferdinando Boero:** These interviews took place at the University of Salento, Lecce, Italy, on October 15, 2014.

151 **a marine biology student collected:** Giorgio Bavestrello, Christian Sommer, and Michele Sara, "Bi-directional Conversion in *Turritopsis nutricula* (Hydrozoa)," *Scientia Marina* 56 (1992): 137–140.

152 **In the lab, Stefano and his collaborators watched the jellyfish morph:** Stefano Piraino et al., "Reversing the Life Cycle: Medusae Transforming into Polyps and Cell Transdifferentiation in *Turritopsis nutricula* (Cnidaria, Hydrozoa)," *Biological Bulletin* 190 (1996): 302–312.

152 **"This process would be hardly more remarkable"**: Stefano Piraino et al., "Reverse Development in Cnidaria," *Canadian Journal of Zoology* 82 (2004): 1748–1754.

153 **proclivity for agelessness might not be constrained:** Jinru He et al., "Life Cycle Reversal in *Aurelia* sp. 1 (Cnidaria, Scyphozoa)," *PLoS One* 10 (2015): e0145314.

154 **That collaborator was:** I interviewed Maria Pia Miglietti on February 18, 2016, at Texas A&M University at Galveston.

154 **the *Turritopsis* living in the waters . . . global range:** Maria Pia Miglietta and Harilaos A. Lessios, "A Silent Invasion," *Biological Invasions* 11 (2009): 825–834.

155 **if we really want to understand the fates of cells:** Alejandro Sánchez Alvarado and Shinya Yamanaka, "Rethinking Differentiation: Stem Cells, Regeneration, and Plasticity," *Cell* 157 (2014): 110–119.

157 **genus *Phialella*, and then dubbed the species *zappai*:** Ferdinando Boero, "Life Cycles of *Phialella zappai* n. sp., *Phialella fragilis* and *Phialella sp.* (Cnidaria, Leptomedusae, Phialellidae) from Central California," *Journal of Natural History* 21 (1987): 465–480.

159 **"Economy and ecology, then"**: Ferdinando Boero, *Economia senza natura: La grande truffa* (Turin: Codice, 2012), pp. 168–169.

159 **erodes the stability of the entire ecosystem:** Gregory L. Britten et al., "Predator Decline Leads to Decreased Stability in a Coastal Fish Community," *Ecology Letters* 17 (2014): 1518–1525.

160 **fishing shifted from individual- and family-owned fishing boats . . . to corporation-owned businesses:** For an accessible account of industrial fishing, see Ted Danson, "Jellyfish Soup," chap. 3 in *Oceana: Our Endangered Oceans and What We Can Do to Save Them* (New York: Rodale, 2011), pp. 75–114.

161 **a gray-ponytailed, black-clad, laser-eyed George Carlin:** George Carlin, "Saving the Planet," https://www.youtube.com/watch?v=7W33HRc1A6c.

12. The Bottom of the Wave

165 **The National Center for Ecological Analysis and Statistics, also known as NCEAS:** I attended the NCEAS meeting in Santa Barbara on October 28–30, 2011.

168 **"the jelly shunt"**: Robert H. Condon et al., "Jellyfish Blooms Result in a Major Microbial Respiratory Sink of Carbon in Marine Systems," *Proceedings of the National Academy of Sciences of the United States of America* 108 (2011): 10225–10230.

168 **jelly falls:** Mario Lebrato et al., "Jelly-Falls Historic and Recent Observations: A Review to Drive Future Research Directions," *Hydrobiologia* 690 (2012): 227–245.

168 **Another type of marine animal, called a salp . . . ocean's vacuum cleaners:** Kelly R. Sutherland, Laurence P. Madin, and Roman Stocker, "Filtration of Submicrometer Particles by Pelagic Tunicates," *Proceedings of the National Academy of Sciences of the United States of America* 34 (2010): 15129–15134.

169 **"they would have reached halfway to the sun":** Colin Schultz, "Massive Salp Swarm Tilts Ocean's Chemical Balance," *EOS Research Spotlight* 93 (2012): 116.

171 **A study of the tissue composition of fish:** Luis Cardona et al., "Massive Consumption of Gelatinous Plankton by Mediterranean Apex Predators," *PLoS One* 7 (2012): e31329.

172 **"the most important predators in the sea":** Daniel Pauly et al., "Jellyfish in Ecosystems, Online Databases, and Ecosystem Models," *Hydrobiologia* 616 (2009): 67–85; Ferdinando Boero et al., "Gelatinous Plankton: Irregularities Rule the World (Sometimes)," *Marine Ecology Progress Series* 356 (2008): 299–310.

172 **In the Chesapeake Bay, jellyfish called sea nettles:** Mary Beth Decker et al., "Predicting the Distribution of the Scyphomedusa *Chrysaora quinquecirrha* in Chesapeake Bay," *Marine Ecology Progress Series* 329 (2007): 99–113.

174 **work analyzing ecosystem-based models of fisheries:** Kelly L. Robinson et al., "Jellyfish, Forage Fish, and the World's Major Fisheries," *Oceanography* 27 (2014): 104–115.

13. Stop Waiting

183 **the giant jellyfish that swarmed the seas of Japan:** As occurs in such claims of superlatives, there's some controversy. In 1865, Harvard's Alex Agassiz wrote of a lion's-mane jellyfish that measured 120 feet, but other lion's manes of that length have not been documented. The siphonophore *Praya dubia* can be longer—200 feet long—though it's a swimming colony of jellyfish, not a single individual. In terms of weight, it's no contest. The giant jellyfish holds the record. For more extremes, see Craig R. McClain et al., "Sizing Ocean Giants: Patterns of Intraspecific Size Variation in Marine Megafauna," *PeerJ* (2015): e715.

184 **And again in 2003 and in 2004:** This article includes an astonishing picture of a dense giant jellyfish bloom surrounding a fishing boat: Masato Kawahara et al., "Unusual Population Explosion of the Giant Jellyfish *Nemopilema nomurai* (Scyphozoa: Rhizostomeae) in East Asian Waters," *Marine Ecology Progress Series* 307 (2006): 161–173.

184 **In 2005 there was a record-setting bloom:** Shin-ichi Uye, "Blooms of the Giant Jellyfish *Nemopilema nomurai*: A Threat to the Fisheries Sustainability of the East Asian Marginal Seas," *Plankton & Benthos Research,* supplement (2008): 125–131.

184 **But 2009 broke the record again:** Shin-ichi Uye, "Human Forcing of the Copepod-Fish-Jellyfish Triangular Trophic Relationship," *Hydrobiologia* 666 (2011): 71–83.

185 **early-detection system:** Akira Okuno et al., "Forecast of the Giant Jellyfish *Nemopilema nomurai* Appearance in the Japan Sea," paper presented at PICES 2011, October 21, 2011, https://www.pices.int/publications/presentations/PICES-2011/2011-BIO/2011-BIO-Day2/BIO-P-Day2-1005-Okuno-7683.pdf.

186 **e-mail from Lucas:** I received this on October 10, 2012.

187 **seed the ocean with baby jellyfish:** Makato Omori and Eiji Nakano, "Jellyfish Fisheries in Southeast Asia," *Hydrobiologica* 451 (2001): 19–26.

187 **from shrimping and traditional fishing to jellyfishing:** The information about jellyfishing is from an e-mail from Shin-ichi Uye on October 10, 2012.

189 **at this meeting I was starstruck:** North Pacific Marine Science Organization, PICES 2012, "Effects of Natural and Anthropogenic Stressors in the North Pacific Ecosystems: Scientific Challenges and Possible Solutions," October 12–21, 2012, Hiroshima.

190 **Masuda said he performed experiments:** Reiji Masuda et al., "Jellyfish as a Predator and Prey of Fishes: Underwater Observations and Rearing Experiments," paper presented at PICES 2012, October 18, 2012.

190 **a small pier installed in Hiroshima Bay:** Masaya Toyokawa et al., "*Aurelia* Swarms Originate from Polyps near the Mouth of a Bay: Evidence from Mikawa Bay and Ise Bay," paper presented at PICES 2012, October 18, 2012.

191 **In Tokyo Bay, moon jellyfish polyps:** Haruto Ishii and Hiromi Shioi, "The Effects of Environmental Light Condition on Strobilation in *Aurelia aurita* Polyps," *Sessile Organisms* 20 (2003): 51–54.

192 **"solar-powered slugs"**: Some research suggests that chloroplast-stealing slugs reap the benefits of photosynthesis, but newer work was unable to find a clear link. Catherine Brahic, "Solar-Powered Sea Slug Harnesses Stolen Plant Genes," *New Scientist*, November 24, 2008, https://www.newscientist.com/article/dn16124-solar-powered-sea-slug-harnesses-stolen-plant-genes; Sarah Zielinski, "Green Sea Slugs Aren't Solar Powered After All," *ScienceNews*, November 20, 2013, https://www.sciencenews.org/blog/wild-things/green-sea-slugs-aren't-solar-powered-after-all.

192 **sea slugs from Barnegat Bay**: Dena Restaino et al., "Molecular Confirmation of Nudibranch Predation on Cnidarian Species in Barnegat Bay, NJ USA," paper presented at the Fifth International Jellyfish Bloom Symposium, Barcelona, May 30–June 3, 2016.

14. Sacred Island

198 **According to the UN, more than 3.1 billion people**: UN Food and Agriculture Organization, *The State of World Fisheries and Aquaculture 2016* (Rome: FAO, 2016), p. 4, http://www.fao.org/3/a-i5555e.pdf.

199 **marine protected areas, or MPAs**: National Marine Protected Areas Center, *Conserving Our Oceans: One Place at a Time*, 2013, http://marineprotectedareas.noaa.gov/pdf/fac/mpas_of_united_states_conserving_oceans_1113.pdf.

199 **In 2016 mega-MPAs were established**: A list of the global commitments toward MPAs in 2016 be found on an Our Ocean webpage, http://ourocean2016.org/commitments#commitments-main, and an assessment of U.S. MPAs can be found in R. Brock, "Representativeness of Marine Protected Areas of the United States," U.S. Department of Commerce, National Oceanic and Atmospheric Administration, National Marine Protected Areas Center, Silver Spring, Maryland, 2015, http://marineprotectedareas.noaa.gov/dataanalysis/mpainventory/rep-report15.pdf.

200 **cordoning off parts . . . actually increases fish yields**: Benjamin S. Halpern and Robert R. Warner, "Matching Marine Reserve Design to Reserve Objectives" *Proceedings of the Royal Society B* 270 (2003): 1871–1878; Crow White et al., "Marine Reserve Effects on Fishery Profit," *Ecology Letters* 11 (2008): 370–379.

201 **foundation of the ocean's food web**: Ted Danson, *Oceana: Our Endangered Oceans and What We Can Do to Save Them* (New York: Rodale, 2011), p. 266.

201 **Lucas received good news from Shin**: Lucas's e-mail containing Shin's message was sent to me on October 20, 2012.

204 **"The paper by the NCEAS group"**: Robert H. Condon et al., "Recurrent Jellyfish Blooms Are a Consequence of Global Oscillations," *Proceedings of the National Academy of Sciences of the United States of America* 110 (2013): 1000–1005.

205 **lukewarm headlines**: Marina Sanz-Martín et al., "Flawed Citation Practices Facilitate the Unsubstantiated Perception of a Global Trend Toward Increased Jellyfish Blooms," *Global Ecology and Biogeography* 25 (2016): 1039–1049.

205 **A 2001 paper**: Claudia E. Mills, "Jellyfish Blooms: Are Populations Increasing Globally in Response to Changing Ocean Conditions?" *Hydrobiologia* 451 (2001): 55–68.

205 **following decade was filled with speculation**: Jenny Purcell, "Jellyfish and Ctenophore Blooms Coincide with Human Proliferation and Environmental Perturbations," *Annual Review of Marine Science* 4 (2012): 209–235; Anthony J. Richardson et al., "The Jellyfish Joyride: Causes, Consequences, and Management Responses to a More Gelatinous Future," *Trends in Ecology & Evolution* 24 (2009): 312–322.

206 **published their study in 2012**: Lucas Brotz et al., "Increasing Jellyfish Populations: Trends in Large Marine Ecosystems," *Hydrobiologia* 690 (2012): 3–20.

15. Stalking the Beast

207 **Lucas and I walked down the gangway**: The material in this chapter is from my notes taken during the trip on October 21–22, 2012.

213 **The creature was optimized**: Masato Kawahara et al., "Unusual Population Explosion of the Giant Jellyfish *Nemopilema nomurai* (Scyphozoa: Rhizostomeae) in East Asian Waters," *Marine Ecology Progress Series* 37 (2006): 161–173.

16. Jellyfish al Dente

216 **"But things have changed recently":** The conversations in this section took place on October 22–23, 2012.

217 **And that may not be the whole story either:** This study lays out the roles that society and politics play in the Ariake Sea's ecosystem health: Momoko Ishikawa and Victor S. Kennedy, "Management of the Oyster Fisheries in Japan's Ariake Sea and Maryland's Chesapeake Bay: A Comparison," *Marine Fisheries Review* 76 (2014): 39.

217 ***Rhopilema esculentum*:** Makoto Omori and Minoru Kitamura, "Taxonomic Review of Three Japanese Species of Edible Jellyfish (Scyphozoa: Rhizostomeae)," *Plankton Biology and Ecology* 51 (2004): 36–51.

17. Jellyfishing

225 **The Yanagawa fish market:** I visited the fish market on October 23, 2012.

225 **That honor goes to the Tsukiji market:** The market has been scheduled for demolition to make way for a highway needed for the 2020 Olympics, but plans are on hold.

228 **worldwide catch of jellyfish:** Lucas Brotz, "Jellyfish Fisheries: A Global Assessment," chap. 10 in *Global Atlas of Marine Fisheries: A Critical Appraisal of Catches and Ecosystem Impacts*, ed. Daniel Pauly and Dirk Zeller (Washington, DC: Island Press: 2016), pp. 110–124.

229 **small pieces of floating plastic:** Delphine Thibault et al., "The Neustonic Environment in the Gulf of Lion: Contribution of Gelatinous Zooplankton and Microplastics," paper presented at the Fifth International Jellyfish Bloom Symposium, Barcelona, May 30–June 3, 2016.

229 **seedlike podocysts:** Masato Kawahara et al., "Bloom or Non-bloom in the Giant Jellyfish *Nemopilema nomurai* (Scyphozoa: Rhizostomeae): Roles of Dormant Podocysts," *Journal of Plankton Research* 35 (2012): 213–217.

234 **"the Bohai Sea":** "Ecosystem Issues and Policy Options Addressing Sustainable Development of China's Ocean and Coast," China Council for International Cooperation on Environment and Development, Annual General Meeting 2010, Reports of Task Force.

237 **a megastorm building on the East Coast:** For an account of the events leading up to Sandy's landfall on the East Coast, see Kathryn Miles, *Superstorm: Nine Days Inside Hurricane Sandy* (New York: Dutton, 2014).

237 **the name Frankenstorm:** Chris Mooney, "Here Comes the Story of No Hurricanes," *Mother Jones*, September 6, 2013, http://www.motherjones.com/environment/2013/09/hurricane-season-ipcc-sandy.

18. Toxic Cocktail

243 **swarms of round white jellyfish:** Bella S. Galil, E. Spanier, and W. W. Ferguson, "The Scyphomedusae of the Mediterranean Coast of Israel, Including Two Lessepsian Migrants New to the Mediterranean," *Zoologische Mededelingen* 64 (1990): 94–105.

245 **"When you remove":** Interview with Enric Sala, National Geographic explorer-in-residence, November 11, 2014.

245 **"the sprinter":** Kimberly Tenggardjaja et al., "Genetics of a Lessepsian Sprinter: The Bluespotted Cornetfish, *Fistularia commersonii*," *Israel Journal of Ecology & Evolution* 59 (2013): 181–185.

245 **All those jellyfish stalled operations:** This short video shows a large bloom of nomadic jellyfish that infiltrated a power plant in Israel: "Israeli Power Plants Threatened by Jellyfish," Agence France-Presse, July 5, 2011, https://www.youtube.com/watch?v=AsUNZBxzddY&feature=player_embedded.

245 **The article explained that the Suez Canal:** Bella S. Galil et al., "'Double Trouble': The Expansion of the Suez Canal and Marine Bioinvasions in the Mediterranean Sea," *Biological Invasions* 17 (2015): 973–976.

247 **When the article was published:** Juli Berwald, "Under the Ships in the Suez Canal," *The New York Times*, November 12, 2014.

248 **Irukandji syndrome:** Lisa-Ann Gershwin et al., "Biology and Ecology of Irukandji Jellyfish (Cnidaria: Cubozoa)," *Advances in Marine Biology* 66 (2013): 1–85.

248 **Hawaiian scientist Angel Yanagihara:** I spoke with Angel Yanagihara by phone on May 26, 2015.

249 ***Alatina* probably lives:** Luciano M. Chiaverano et al., "Long-Term Fluctuations in Circalunar Beach Aggregations of the Box Jellyfish *Alatina moseri* in Hawaii, with Links to Environmental Variability," *PLoS One* 8 (2013): e77039.

250 **In 2001, Angel identified a porin:** John J. Chung et al., "Partial Purification and Characterization of a Hemolysin (CAH1) from Hawaiian Box Jellyfish (*Carybdea alata*) Venom," *Toxicon* 39 (2001): 981–990.

251 **Researching old cures:** Angel A. Yanagihara and Ralph V. Shohet, "Cubozoan Venom–Induced Cardiovascular Collapse Is Caused by Hyperkalemia and Prevented by Zinc Gluconate in Mice," *PLoS One* 7 (2012): e51368.

251 **champion endurance swimmer Diana Nyad:** For an engrossing account of Nyad's quest, see her memoir *Find a Way: The Inspiring Story of One Woman's Pursuit of a Lifelong Dream* (New York: Vintage, 2015).

252 **"If you are going":** I spoke with Diana Nyad by phone on September 9, 2016.

252 **Steve Munatones, who was the official observer:** I spoke to Steve Munatones by phone on September 9, 2016.

254 **The formulation worked:** Additional funding has allowed Angel to develop a commercially available version of the cream, used by combat divers and the general public, called Sting No More (http://stingnomore.com).

255 **She and her colleagues had looked at the ingredients:** Tamar Rachamim et al., "The Dynamically Evolving Nematocyst Content of an Anthozoan, a Scyphozoan, and a Hydrozoan," *Molecular Biology and Evolution* 32 (2015): 740–753.

256 **"there are large gaps":** Sophie Badré, "Bioactive Toxins from Stinging Jellyfish," *Toxicon* 91 (2014): 114–125.

19. Sting Block

259 **the idea for his sting block:** I spoke with Amit Lotan at the Safe Sea headquarters near Tiberias on July 14, 2015.

259 **The skin of the capsule:** Anna Beckmann et al., "A Fast Recoiling Silk-like Elastomer Facilitates Nanosecond Nematocyst Discharge," *BioMed Central Biology* 13 (2015): doi: 10.1186/s12915-014-0113-1.

260 **150 atmospheres:** Suat Özbek, Prakash G. Balasubramanian, and Thomas W. Holstein, "Cnidocyst Structure and the Biomechanics of Discharge," *Toxicon* 54 (2009): 1038–1045.

261 **astonishing number of riffs:** Carina Östman, "A Guideline to Nematocyst Nomenclature and Classification, and Some Notes on the Systematic Value of Nematocysts," *Scientia Marina* 64 (2000): 31–46.

261 **I contacted Glen:** I spoke to Glen Watson by phone on May 12, 2016.

261 **what an anemone would "smell":** Glen M. Watson and David A. Hessinger, "Cnidocyte Mechanoreceptors Are Tuned to the Movements of Swimming Prey by Chemoreceptors," *Science* 243 (1989): 1589–1591.

262 **multiple passwords on your phone:** Glen M. Watson and Renee R. Hudson, "Frequency and Amplitude Tuning of Nematocyst Discharge by Proline," *Journal of Experimental Biology* 268 (1994): 177–185.

263 **arm it with a high-tech trigger:** Pei-Ciao Tang, Karen Müller Smith and Glen M. Watson, "Repair of Traumatized Mammalian Hair Cells via Sea Anemone Repair Proteins," *Journal of Experimental Biology* 219 (2016): 2265–2270.

263 **ingredients in his sting block:** Alexa Boer Kimball et al., "Efficacy of a Jellyfish Sting Inhibitor in Preventing Jellyfish Stings in Normal Volunteers," *Wilderness & Environmental Medicine* 15 (2004): 102–108.

264 **a popular Spanish television show:** "El Hombre de Negro se mete en una pecera llena de medusas," *El Hormiguero* 3.0, YouTube, July 7, 2015, https://www.youtube.com/watch?v=wO2sNfkvpLY&list=PL3724FB368BDC886C&index=9.

265 **clinical trials in Belize, Florida, and California:** David R. Boulware, "A Randomized, Controlled Field Trial for the Prevention of Jellyfish Stings with a Topical Sting Inhibitor," *Journal of Travel*

Medicine 13 (2006): 166–171; Kimball et al., "Efficacy of a Jellyfish Sting Inhibitor in Preventing Jellyfish Stings in Normal Volunteers."

20. In Medusa's Blood

271 **I called the director of the institute:** I spoke with Maoz Fine on July 21, 2015, in Eilat.

272 **The process is known as bleaching:** Ove Hoegh-Guldberg, "Climate Change, Coral Bleaching and the Future of the World's Coral Reefs," *Marine and Freshwater Research* 50 (1999): 839–866.

272 **Great Barrier Reef:** "Life and Death After Great Barrier Reef Bleaching," ARC Centre of Excellence, Coral Reef Studies, media release, November 29, 2016.

272 **The International Society for Reef Studies:** "Climate Change Threatens the Survival of Coral Reefs," ISRS Consensus Statement on Climate Change and Coral Bleaching, October 2015, Conference of the Parties to the United Nations Framework Convention on Climate Change, 21st session, Paris, December 2015, http://coralreefs.org/wp-content/uploads/2014/03/ISRS-Consensus-Statement-on-Coral-Bleaching-Climate-Change-FINAL-14Oct2015-HR.pdf.

272 **something special about these corals:** Maoz Fine, Hezi Gildor, and Amatzia Genin, "A Coral Reef Refuge in the Red Sea," *Global Change Biology* 19 (2013): 3640–3647.

274 **The corals lost their skeletons:** Maoz Fine and Dan Tchernov, "Scleractinian Coral Species Survive and Recover from Decalcification," *Science* 315 (2007): 1811.

275 **one study from Spain:** María Algueró-Muñiz et al., "Withstanding Multiple Stressors: Ephyrae of the Moon Jellyfish (*Aurelia aurita*, Scyphozoa) in a High-Temperature, High-CO_2 and Low-Oxygen Environment," *Marine Biology* 163 (2016): 185.

275 **scientists in Japan and England found similar results:** Oliver Tills et al., "Reduced pH Affects Pulsing Behaviour and Body Size in Ephyrae of the Moon Jellyfish, *Aurelia aurita*," *Journal of Experimental Marine Biology and Ecology* 480 (2016): 54–61; Takashi Kikkawa et al., "Swimming Inhibition by Elevated pCO_2 in Ephyrae of the Scyphozoan Jellyfish *Aurelia*," *Plankon & Benthos Research* 5 (2010): 119–122.

278 **early for my meeting with Yossi Loya:** I spoke with Yossi Loya on July 22, 2015, at Tel Aviv University.

279 **The quantitative tools that Yossi developed:** Yossi Loya, "Plotless and Transect Methods," *Monographs on Oceanic Methodology*, ed. D. R. Stoddart and R. E. Johannes (Paris: UNESCO Press, 1978), pp. 197–218.

279 **affecting the reproduction:** Yossi Loya and Baruch Rinkevich, "Abortion Effects in Corals Induced by Oil Pollution," *Marine Ecology Progress Series* 1 (1979): 77–80.

280 **But the reef's troubles:** Yossi Loya, "The Coral Reefs of Eilat—Past, Present and Future: Three Decades of Coral Community Structure Studies," *Coral Reef Health and Disease*, ed. E. Rosenberg and Y. Loya (Berlin: Springer-Verlag, 2004), pp. 1–34.

280 **The battle to evict the fish farms:** Yossi Loya, "How to Influence Environmental Decision Makers? The Case of Eilat (Red Sea) Coral Reefs," *Journal of Experimental Marine Biology and Ecology* 344 (2007): 35–53.

280 **Yossi showed me a graph:** Tom Shlesinger and Yossi Loya, "Recruitment, Mortality, and Resilience Potential of Scleractinian Corals at Eilat, Red Sea," *Coral Reefs* 35 (2016): 1357–1368, figure 9.

280 **environmental stewardship and regulation:** Israel Ministry of Environmental Protection, "Gulf of Eilat," updated December 5, 2014, http://www.sviva.gov.il/English/env_topics/MajorBodiesOfWater/Pages/GulfOfEilat.aspx?WebId=Gulf_of_Eilat.

21. Party Like a Jellyfish

283 **meet for dinner in Tel Aviv:** I spoke with Bella Galil on July 23, 2015, in Tel Aviv.

285 **more carbon dioxide than any other state in the country:** "Energy-Related Carbon Dioxide Emissions at the State Level, 2000–2014," U.S. Energy Information Administration, http://www.eia.gov/environment/emissions/state/analysis.

285 **seventh-largest emissions output in the world:** "Each Country's Share of CO_2 Emissions," Union of Concerned Scientists. The data in this list were taken from the U.S. Energy Information Agency but reformatted to rank emissions by country. http://www.ucsusa.org/global_warming/science_and_impacts/science/each-countrys-share-of-co2.html#.V8XSVGVlmRs.

285 **local group of citizen lobbyists:** I joined Citizens' Climate Lobby (https://citizensclimatelobby.org), which supports a carbon fee and dividend. In this mechanism, carbon is taxed at the point of extraction, and the revenues are redistributed to the taxpayers to offset the increased price of electricity to the consumer, while leaving the details of the shift away from carbon emissions to market forces rather than regulation. The group is nonpartisan, with support from both the left and the right. Such inclusivity is key. We need as many people working together as possible.

286 **the opportunistic ones:** Bella Galil, "Truth and Consequences: The Bioinvasion of the Mediterranean Sea," *Integrative Zoology* 7 (2012): 299–311.

286 **marine protected areas were being taken over:** M. Otero et al., *Monitoring Marine Invasive Species in Mediterranean Marine Protected Areas (MPAs): A Strategy and Practical Guide for Managers* (Málaga, Spain: IUCN, 2013).

289 **a man rose like a sea monster:** My meeting with Moti Mendelson was on July 24, 2015, at Kiryat Yam in Haifa.

22. Bloom

295 **the Mar Menor, or the Little Sea:** Héctor M. Conesa and Francisco J. Jiménez-Cárceles, "The Mar Menor Lagoon (SE Spain): A Singular Natural Ecosystem Threatened by Human Activities," *Marine Pollution Bulletin* 54 (2007): 839–849.

296 **a report by the Regional Ministry of Agriculture, Water, and the Environment:** Región de Murcia Consejería de Agricultura, Agua y Medio Ambiente, "Seguimiento biológico de las poblaciones de medusas en el Mar Menor, Campañas 2000–2003," report provided by Cristina Mena.

296 **such jellyfish-control programs:** Cathy H. Lucas, Stefan Gelcich, and Shin-ichi Uye, "Living with Jellyfish: Management and Adaptation Strategies," *Jellyfish Blooms*, ed. Kylie A. Pitt and Cathy H. Lucas (Dordrecht: Springer Science + Business Media, 2014), pp. 129–150; G. Rodriguez, G. Clarindo, and L. McKnight, "Jellyfish Outbreaks in Coastal City Beaches from a Management Perspective," *Coastal Cities and Their Sustainable Future* (Southampton, England: WIT Press, 2015), pp. 277–288.

296 **San Pedro del Pinatar:** I met with Cristina Mena and Victor Moreno on June 4, 2016.

298 **Medusas team:** Much information about the program came from talking with Victor and Cristina. Also see "Control y estudio de las poblaciones de medusas en el Mar Menor," Comunidad Autónoma de la Región de Murcia, https://www.carm.es/web/pagina?IDCONTENIDO=908&IDTIPO=11&RASTRO=c494$m1268.

298 **400,000 euros a year:** "Mar Menor Anti-Jellyfish Nets Ready to Be Put in Place," *Murcia Today*, May 13, 2016, http://murciatoday.com/mar-menor-anti_jellyfish-nets-ready-to-be-put-in-place_30111-a.html#leftcol.

298 **Fifth International Jellyfish Blooms Symposium:** See http://www.jellyfishbloom2016.com.

300 **Monty Graham opened the proceedings:** William M. Graham, "Jellyfish and Human Well-Being," paper presented at the Fifth International Jellyfish Blooms Symposium, Barcelona, May 30–June 3, 2016.

301 **In her plenary talk the next day:** Jennifer E. Purcell, "Large-Scale and Long-Term Perspectives on Jellyfish Research," paper presented at the Fifth International Jellyfish Blooms Symposium, Barcelona, May 30–June 3, 2016.

301 **reports to jellyfish watch websites:** Based in California, jellywatch.org collects and posts jellyfish sightings from around the world and also provides links to local citizen science projects that collect information about jellyfish.

302 **"Obama has spoken of a nuclear-free future":** Shin-ichi Uye, "Forecast for the Annual Bloom Intensity of the Giant Jellyfish *Nemopilema nomurai*: 10-Year Monitoring Using Ships of Opportunity," paper presented at the Fifth International Jellyfish Blooms Symposium, Barcelona, May 30–June 3, 2016.

302 **According to a 2012 study:** Robert H. Condon et al., "Questioning the Rise of Gelatinous Zooplankton in the World's Oceans," *BioScience* 62 (2012): 160–169.

303 **As evidenced by the topics covered at the meeting:** For a list of sessions and descriptions at the meeting, see http://www.jellyfishbloom2016.com/sessions.

Index

Page numbers in italics refer to illustrations.

Abalone, 41, 226
Acidification, 8, 10, 12, 178, 223, 274–76, 285
Acre (Israel), 289
Acropora, 278
ADHD (attention deficit hyperactivity disorder), 190
Adriatic Sea, 30, 74
"Adventure of the Lion's Mane, The" (Doyle), 12
Aegean Sea, 30, 56, 76
Aequorea forskalea, 32
Aequorea victoria, 111, 187
Aequorin, 114
Africa, 279
Agassiz, Louis, 11
Aglow in the Dark (Pieribone and Gruber), 114
Agriculture, 15, 160
Alabama, 22, 37–39, 53, 58, 165, 167, 298
Alaska, 36, 77
Alatina, 249–50
Albert I, Prince of Monaco, 11
Algae, 9, 51, 55, 61, 65, 234, 270, 272–74
Alligator turtles, 194
Alzheimer's disease, 47, 117, 189
Anaphylaxis, 11
Anchovy fisheries, 28–30, 32, 159, 175, 226
Andon, Alex, 53–57, 62, 95, 142
Anglerfish, 104
Antarctica, 31
Antioxidants, 50, 52
Anti-Semitism, 73–74
Apoaequorin, 114
Aquaculture, 15, 26–27, 63, 158, 187, 217, 231, 280, 303
Aramaki, Katsueda, 218, 220–22, 225, 230–33
Aramaki, Teruyuki, 218–20
Architeuthis, 106
Arctic, 13, 14
Argentina, 73
Ariake Sea, 186, 187, 216–19, 229, 233–36, 238
Aristotle, 43
Art Forms in Nature (Haeckel), 12
Art Nouveau, 12, 71, 165
Aswan Dam, 244
Athena, 7
Atlantic Ocean, 9, 27, 30, 31, 73, 75, 154, 199
Atolla wyvillei, 100, 105
Atomic Bomb Dome (Hiroshima), 3
Audubon Society, 199
Aurelia, 39, 76, 79
Austin (Texas), 6, 117, 166, 176–77, 199, 221
  University of Texas at, 222
Australia, 25, 26, 30, 64, 77, 168, 169, 253, 279, 296
Axolotls, 117
Azores, 73
Azov, Sea of, 30

Bahamas, 251
Baja California, 101
Bakalar, David, 94
Ballast water, invasive species in, 9, 27, 30–31, 154
Baltic Sea, 30
Barcelona, 298–302
Barnacles, 65, 171, 198
Barnegat Bay (New Jersey), 192
Barrel jellyfish, 51, 64, 65, 264, 296
Bayha, Keith, 36–37, 70–71, 73, 76–79, 85, 299

Beagle (ship), 73
Belize, 265
Benguela Current, 31
Bering Sea, 173, 174, 236
Bermuda, 168, 175, 176, 185
Beroe ovata, 29
Big Balling (TV show), 46
Big Cobb (boat), 44
Big red jellyfish, 99–100
Bioluminescence, 14, 29, 102–6, 109–13, 115–16, 226, 303
Bitter Lakes, 244, 246
Black Sea, 9, 27–30, 33, 81, 154
Blood belly comb jellyfish, 99
Blooms, 14–15, 32–34, 77, 187, 205, 224, 245
  absence of, 224, 229–30, 281, 283
  of comb jellyfish, 9, 30, 112–13
  conferences on, 22, 298–303
  equipment clogs caused by, 25, 243, 258
  of giant jellyfish, 184–86, 224
  impacts of, 8–10, 25–26, 167–72, 189, 235, 296
  jellyfishing catch from, 45
  of phytoplankton, 101, 217
  of pollution-caused algae, 234
  polyps proliferating into, 64, 230
  swimming in, 276, 284, 288, 290–93
Blue dragon slug, 192
Bluefin tuna, 171, 201
Blue-water diving, 13, 94, 96, 175, 176
Bodega Bay (California), 157
Boero, Ferdinando, 150, 156–62, 198, 236, 245–46
Bohai Sea, 234, 235
Bosphorus, 27
Boston Public Library, 97
Botany of Desire, The (Pollan), 118
Box jellyfish, 64, 78, 79, 123–26, 247–55, 265, 299
BP, 37. *See also* Deepwater Horizon oil spill
Brave New World (Huxley), 11
Brazil, 73, 107
Brisbane, 25
British Columbia, University of, 63, 189
Brotz, Lucas, 63, 184–86, 189, 193–95, 198, 201–17, 219–24, 227–34, 299
Buoyancy, 58, 68, 92, 98, 269
Butterfish, 226
Bycatch, 45, 228
Cabbage head jellyfish. *See* Cannonball jellyfish
Caesarea (Israel), 283
California, 79, 156–57, 168, 174, 265
  agriculture in, 57
  carbon emissions in, 285
  desalination in, 25
  marine protected areas in, 200
  University of, Merced, 77
  See also specific cities, regions, and natural features
California Institute of Technology, 91
Cambridge, University of, 68, 132, 133, 139
Canada, 68
Canary Islands, 73
Cancer, 117, 118, 154, 190
Cannonball jellyfish, 44, 50, 219, 220
Cape Canaveral (Florida), 128–29
Cape Cod (Massachusetts), 84–94
Carbon emissions, 6, 8, 70, 170, 285
Caribbean, 279
Carlin, George, 157, 161–62
Carson, Rachel, 300
Caspian Sea, 30
Chalfie, Martin, 116, 118–19
Channel Islands (California), 101
Charismatic megafauna, 24
Chesapeake Bay, 10, 59, 172
Chiaverano, Luciano, 79
Chicago, University of, 84
Chile, 199
China, 54, 79, 213, 228, 299
  Academy of Sciences of, 15
  carbon emissions in, 285
  giant jellyfish in, 229
  jellyfish cuisine in, 43, 47, 48, 187, 188, 219
  marine environment acidification in, 234
  research on medical uses of jellyfish in, 50
  sting treatment developed in, 253
Chironex fleckeri, 26, 248
Christianity, 145
Chrysaora, 7
  Chrysaora fulgida, 6, 32
  See also Sea nettles
Chuang, Terry, 46
Clams, 5, 106, 127, 172, 216, 218, 233, 278
Clare Island (Ireland), 26
Clean Air Act (1970), 301
Clean Water Act (1972), 301
Climate change, 10, 15, 58, 63, 178, 206, 271
  coral reefs threatened by, 273–74, 279, 281
  evolution and, 69–70

impact on oceans of, 178, 271
Monterey Bay Aquarium exhibit on, 60–61
power of storms increasing due to, 237–38
Cnidae, 11
Cnidaria, 11, 81, 255–56, 258–59
Coastal development, 15, 63, 206, 223, 234, 279, 302
Coelenterates, 10–11
Colin, Sean, 87–90, 92–93
Collagen, 49–50, 259
Columbia space shuttle, 129, 275
Columbia University, 116
Comb jellyfish, 10–11, 13, 27–30, 70, 99, 103, 303
in ballast water, 9, 27, 154
in Black Sea, 27–30, 113
evolution of, 70, 80–81
predators of, 29–30, 172–73
transparency of, 98
worldwide invasions of, 15, 30
Compass jellyfish, 6, 32–33
Condon, Rob, 36, 165, 175, 176
Congress, U.S., 285
Copenhagen, University of, 123
Copepods, 103, 169, 233
Corals, 4–5, 8, 10–11, 68, 81, 199–200, 261, 270–80
See also Reefs
Cornell University, 286
Costa Brava (Spain), 26
Costello, Jack, 87–90, 92–93
Cotylorhiza, 290
Cotylorhiza tuberculata, 296
Cousteau, Jacques, 276
Crabs, 8–9, 12, 46, 127, 170, 217, 226
Ctenophora, 11
Crimea, 27
Croatia, 30
Dalmatian coast of, 71, 74, 75
Cuba, 251, 254–55
Cyaneidae, 72
Cyro droid, 90–91

Dallas, University of Texas at, 90
Dalmatian coast, 71, 74, 75
Darwin, Charles, 11, 71, 73, 132, 133, 142–44, 157–59
Darwin, Francis, 132
Dauphin Island Sea Lab (Alabama), 22, 58, 70, 73, 223
Dawkins, Richard, 118
Dawson, Michael, 77–79
Daytona (Florida), 25
DDT, 301
Deep End Dining (blog), 48–49
Deepwater Horizon oil spill, 37–39
Denmark, 300
Denver Museum of Nature & Science, 67–69
Desalination plants, 25
Discovery Channel, 89, 107, 194
DNA, 57, 65, 71, 77, 80–81, 116, 192
Dolphins, 22, 92, 174, 292
Doyle, Arthur Conan, 12
Drymonema
Drymonema dalmatina, 74–76
Drymonema larsoni, 78
Drymonematidae, 78
Duke University, 53, 97

Earle, Sylvia, 158
Eastern Virginia Medical School, 128, 129
Echinoderms. *See* Sea urchins
Echizen kurage, 183–84. *See also Nemopilema nomurai*
Economy Without Nature: The Great Swindle (Boero), 158
Ecosystems, 13, 106, 165, 167–75, 178, 206, 223, 301
of coral reefs, 5, 272
human-caused degradation of, 15, 27, 29, 160–61, 217
of hydrothermal vents, 69
impact of jellyfish blooms on, 8–10, 167–70, 235
invasive species in, 31, 33, 245–46, 286–87
protection of, 199–201, 246, 287
predators' importance in stability of, 159–60, 171–73
reprogramming of cells in, 154–56
Eel Pond (Woods Hole), 86
Egypt, 247, 273
Eilat (Israel), 4, 266, 267, 270–74, 278–81
El Niño/La Niña pattern, 15, 32
Encyclopædia Britannica, 143
Endangered species, 76, 78, 200
England, 132, 275
Ephyra (sea nymph), 56

Ephyrae, *ix*, 61, *181*, 191, 229
artificial, in Cyro, 91
budded, 56–57, 62, 183
reverse aging of, 152
in space experiments, 129–30
Esposito, Gennaro, 51
Ethiopia, 266
European Parliament, 247
Evolution, 69, 80–82, 112, 118, 121, 125, 136, 259, 303
Darwin's theory of, 11, 142

Fernez, Maurice, 276
Finding Nemo (movie), 99
Fine, Maoz, 271–76
Fire coral, 4, 270, 278
Fish farms. *See* Aquaculture
Fish and Wildlife Service, U.S., 78
Fisheries, 73, 158, 161, 171, 301, 303
benefits of marine protected areas for, 200
decline and collapse of, 9, 24, 27–33, 234–35
jellyfish, 44–46, 64, 187, 228
models of, 173, 174
Florida, 25, 31, 46, 128–29, 154, 199, 251, 254–55, 265. *See also specific cities and regions*
Flotation. *See* Buoyancy
Fluorescence, 115–18
Foça (Turkey), 76
Focke, Wilhelm Olbers, 143–45
Fossil fuels, 8, 15, 24, 161, 170, 178–79, 285
France, 253, 299
Friday Harbor Labs, 111
Frogfish, 226
Fukuoka (Japan), 201, 202
Fungia, 278

Gale, Wynn, 44–46, 220
Galil, Bella, 243–35, 246, 247, 283–84, 286–88, 290
Galilee, Sea of, 258
Galveston, 154
Garm, Anders, 123–26
Gelata, 11, 12, 14
Gemmell, Brad, 90
Genoa, 151, 157
University of, 159
Georgia, 44–46, 219, 220
Germany, 71, 118, 152, 285
Nazi, 73–75
Giant jellyfish, 13, 31, 188, 208–14, 228–32
blooms of, 9, 26, 184–85, 224
bycatch of, 193, 228
culinary, 43–44, 219–21
deterioration out of water, 213–14, 232
diving with, 194–95, 201, 211
names in different languages for, 183
polyps of, 64, 184, 230
in *Rise of the Jellyfish*, 58, 194
seeding areas for, 229
sightings of, 201–2, 210
stings of, 15, 213, 253
weight of, 39, 184, 187, 213, 221
See also Sea nettles
Gibraltar, 245
Glaucus atlanticus (blue dragon), 192
Global warming, 8, 10, 15, 285. *See also* Climate change
Gold Coast (Australia), 26
Golden Island International, 46–48
Graham, Monty, 22–24, 33–36, 40, 195, 298–302
at International Jellyfish Symposiums, 189, 194, 298–302
on jellyfish populations, 34–35, 223
in National Center for Ecological Analysis and Synthesis project, 35, 36, 165, 167, 170, 172–75
on negative impacts of jellyfish, 24–26, 33
in *Rise of the Jellyfish*, 58–60, 89
Great Barrier Reef (Australia), 272
Greeks, ancient, 6–7, 56, 254
Green fluorescent protein (GFP), 115–18
Greenland, 73, 237
Gruber, David, 114
Gulf of Maine, 103
Gulf of Mexico, 14, 27, 77, 79, 174
Deepwater Horizon explosion and oil spill, 37–39
electronic bioluminescent jellyfish (e-jelly) deployed in, 105–6
jellyfish blooms in, 26, 64
jellyfishery in, 44
Gulf Shores (Alabama), 22

Haddock, Steve, 10, 12
Haeckel, Ernst, 12, 71–74, 158, 165
Hagadorn, James "Whitey," 67–69

Hagfish, 168
Haifa, 246, 284, 287–88
  University of, 255
Hake, 32
Haootia quadriformis, 68–69
Harbison, Richard, 97
Harper, April, 46–48
Harvard University, 11, 91
Hawaii, 25, 29, 199, 248–50
He, Jinru, 153
Hermosa Beach (California), 101–2
Hirohito, Emperor of Japan, 11
Hiroshima, 3, 110, 184–85, 188–95, 197, 198, 201, 216, 233, 302
History of Animals, The (Aristotle), 43
HIV (human immunodeficiency virus), 117
Homer, 76
How to Keep Jellyfish in Aquariums (Widmer), 60
Hungarian National Museum, 74, 75
Huxley, Thomas, 11
Hybridization, 142–43
Hydra polyp, 255
Hydromedusa, 151–52
Hydrothermal vents, 69
Hydrozoans, 11, 78, 79, 156
Hyperiid amphipods, 13

India, 228, 300
Indian Ocean, 30, 31, 244, 273
Indonesia, 26, 228, 300
Inland Sea (Japan), 198
International Jellyfish Blooms Symposiums, 298–303
International Society for Reef Studies, 272
Interuniversity Institute for Marine Sciences, 270–76
  Red Sea Simulator, 271–73
Invasive species, 26–27, 29, 63, 229, 244, 247, 286
Ionian Sea, 51
Iran, 118
Ireland, 14, 26
Irukandji syndrome, 248–49
Isahaya Bay (Japan), 217
Israel, 255–81, 283–84, 299
  clogs of power plants and desalination plants in, 24, 25
  coral reefs of, 266–81
  jellyfish blooms off, 30, 276, 283–84, 287–93
  sting research in, 253, 255–65
  See also Jerusalem; Tel Aviv
Italy, 30, 50, 147, 150–52, 154, 190, 245, 299. *See also* Genoa; Leccce; Rome; Salento, University of
Itsukushima Island (Japan), 195, 197–98
Izuhara (Japan). *See* Tsushima Island

Japan, 107, 201–3, 207–35, 245, 267, 299, 300
  environmental research in, 275, 269
  fish markets in, 225–28
  Fisheries Agency, 184, 185
  giant jellyfish of, 9, 15, 58, 64, 183–88, 212–14, 219–20, 224
  jellyfish blooms off, 14, 26, 224, 229
  jellyfish cuisine in, 48, 219–22
  jellyfishing in, 211–14, 216, 218–22, 225, 230–32
  power outages caused by jellyfish in, 24, 33, 79
  Sea of, 43, 186, 193, 201
  seaweed farming in, 217–18
  Turritopsis DNA research in, 154–55
  in World War II, 110–11
  See also specific cities, islands, and regions
Jellyball. *See* Cannonball jellyfish
Jelly-fish, Star-fish and Sea-Urchins (Romanes), xi, 134
Jellyfish Act (1966), 10
Jellyfish Art, 53–54, 95, 142
Jellyfish Ecological Database Initiative (JEDI), 35
Jerusalem, 243, 257–58, 265, 267, 290, 293
Jena (Germany), 71, 73
Jernigan, Tamara (Tammy), 129
Jews, and anti-Semitism, 73–74
Johnsen, Sönke, 97–100, 102, 107
Johnson, Frank, 111
Jordan, 268, 271
Jordan River, 265

Kelp, 158, 172, 191
Key West (Florida), 254–55
Keystone species, 172
King, Rodney, 100
Kiryat Yam (Israel), 284
Korea and Korean Peninsula, 79, 184, 194, 207, 213, 253

Kreisels, 54, 55, 60, 61
Kyushu Island (Japan), 201, 216

Larson, Rob, 75–76, 78
Larvae
  jellyfish. *See* Planulae
  oyster, 172–73
Leatherback turtles, 226
Lecce (Italy), 152
Leeuwenhoek, Anton van, 119
Leiden (Netherlands), 73
Leone, Antonella, 50–51, 190, 299
Lessepsian migrants, 244
Life in Moving Fluids (Vogel), 85–87
Liittschwager, David, 8
Lingulodinium polyedrum, 102
Linnaeus, Carl, 109
Lion's mane jellyfish, 12
Lobsters, 8–9, 12, 13, 178
Loch Duart (Scotland), 26
London, 44
Los Angeles, 100–101, 136–37, 147, 177
Lotan, Amit, 258–61, 263–66, 276, 299
Lotan, Tamar, 255, 258, 265
Loya, Yossi, 278–80
Luciferin, 111–15
Luzon Island (Philippines), 24

MacArthur Genius Grant, 107
Mackerel, 32, 190, 226, 244
Magueyes (Puerto Rico), 75
Maine, Gulf of, 103
Majorca, 296
Malaysia, 228
Malta, 296
Manhattan, 251
Mar Menor (Little Sea), 9, 295–98
Marianas Trench, 107
Marine Harvest, 26
Marine protected areas (MPAs), 199–200, 286–87
Marlin, 201
Marmara, Sea of, 38
Maslow, Abraham, 300
Massachusetts, 84–94
Massachusetts Institute of Technology (MIT), 84
Mastigias papua, 61
Masuda, Reiji, 190, 191
Mauve stinger, 26, 246
Mayer, Alfred Goldsborough, 43
Mediterranean, 76, 173, 247, 253, 295–98
  blooms in, 26, 64, 243–45
  and invasive species introduced via ship traffic, 27, 30, 31, 74
  jellyfish predators in, 171, 292
Medusa (mythological monster), 6–7, 266, 299
Medusae, *ix*, 6–8, 13, 27, 75–76, 83, 173, 186, *241*, 293
  acidification and, 276
  in aquaria, 61
  bioluminescence of, 109–10
  blooms of, 32–34, 184
  culinary, 43, 187–88
  evolution of, 67–69
  polyp production of, 63–65, 157, 183, 191, 229–30
  reproductive process of, 28, 55–57, 72, 169, 203–4
  reverse aging of, 151–55
  Robojelly of, 87–88
  stings of, 264–65, 291–92
Medusae of the World (Mayer), 43
Medusas program, 296–98
Megafauna, charismatic, 24
Mena, Cristina, 296–98
Mendel, Gregor, 143–45
Mendelson, Moti, 287, 289–93
Merced, University of California at, 77
Mesoglea, 58–59, 89–90, 98–99
Mexico, 199, 228. *See also* Baja California; Gulf of Mexico
Miglietta, Maria Pia, 154–55
Milan Expo (2016), 51
Millepora, 278
Miyajima (Shrine Island, Japan), 195, 197–98, 267
Mnemiopsis leidyi, 27–30, 81, 154. *See also* Comb jellies
Mobile Bay (Alabama), 22
Mongooses, 29
Monk seals, 76
Monterey (California), 101
Monterey Bay Aquarium, 54, 57, 60–62, 152, 299
Monterey Bay Research Institute, 10
Moon jellyfish, 39, 58, 76, 114, 296, 300
  in aquaria, 4, 54, 96, 100, 122, 132
  culinary, 220
  efficiency of motion of, 90, 92

flotation of, 59
medusae of, 65, 153
new species of, 78–79
polyps of, 54–55, 62–64, 190–91
Robojelly of, 88
sensory structures of, 122–24, 128, 129, 138
toxins of, 255
Moreno, Victor, 297
Munatones, Steve, 252
Mussels, 106, 216

Nagasaki, 110
Nagoya University, 110
Namibia, 9, 31–33
Naples (Italy) Zoological Station, 73
NASA (National Aeronautics and Space Administration), 107, 128–29
National Center for Ecological Analysis and Synthesis (NCEAS), 35, 36, 165–66, 204–6
National Geographic, 6, 8–9
National Research Council, Institute of Sciences of Food Production, 50
National Science Foundation, 25
Natural selection, 142, 259
Nature (journal), 124
Navy, U.S., 87
Nazis, 73–75
Nemerteans, 73
Nemertes (sea nymph), 73
Nemopilema nomurai (Nomura's jellyfish), 58, 183, 186, 202, 230
Nereids (sea nymphs), 73
Netherlands, 65, 73
Neurotransmitters, 114, 139–40
New Guinea, 77
New Jersey, 296
New Panamax ships, 30
New York, 31
State University, Stony Brook, 279
New York Times, The, 247
Newfoundland, 68
Nicaragua, 31
Nile River, 244
Nobel Prize, 11, 94, 118–19, 139, 155
Nomura's jellyfish. *See Echizen kurage*; *Nemopilema nomurai*
North Carolina, 199
North Sea, 30, 65
Northern Ireland, 26
Norway, 14, 77, 300
Norwegian Sea, 173
Nudibranchs, 191–92
Nyad, Diana, 251–55, 299

Obama, Barack, 199, 302
Obelia, 229
Ocean exploration versus space exploration, 107
Ocean Research & Conservation Association (ORCA), 107
Oceana (nonprofit), 45
Oceanus (god), 56
Octopus, 8, 124, 226, 297
Oil spills, 37–38, 279–80
Oman, 25
On the Origin of Species (Darwin), 132, 158
Oregon, 78, 199
Orkney Islands (Scotland), 26–27
Ostracod, 103, 110, 111, 113
Overfishing, 15, 27, 32, 63, 80, 187, 201, 206, 223, 302
Ovid, 7
Oysters, 134, 172–73, 198, 233

Pacific Ocean, 31, 69, 79, 101, 111, 172, 199, 238, 247–48
Palau, 199
Panama, 154
Panama Canal, 30
Parkinson's disease, 154
Patry, Wyatt, 60–62, 299
Pegasus, 7
Pelagia, 70
Pelican Island National Wildlife Refuge (Florida), 199
Perseus, 7, 266
Phialella zappai, 157
Philippines, 9, 24
Photosynthesis, 61, 101, 192, 217, 272
Phyllorhiza, 290
Phytoplankton, 32, 101–2, 104, 166, 168–69, 216–17
Pieribone, Vincent, 114
Piraino, Stefano, 150–56, 236, 245–46, 299
Pitcairn Islands, 199
Pitt, Kylie, 168–70

Plankton, 5, 12, 57, 69, 73, 101, 104, 169, 171–72, 176, 217
vertical migration of, 123, 252
See also Phytoplankton; Zooplankton
Planulae, *ix*, *1*, 55, 62, 123, 152, 233
Pliny the Elder, 109
Pocillopora, 278
Podocysts, 56, 65, 229–30
Pollan, Michael, 118
Polydectes (mythological king), 266
Polypodium hydriforme, 68
Polyps, *ix*, 16, *19*, 67, 173, 184, 190, 233, 301
budding of, 56–57, 62, 151, 183
coral, 278
influence of climate and acidification on, 9, 33, 274, 276
predators of, 191–93
physical structure of, 55, 151
podocysts produced by, 229–30
predators of, 191–93
reproductive process of, 56, 63, 203
responses to light of, 123
reverse aging of, 151–55
solid surface habitats for, 63–65, 183, 190–91, 229
in space mission experiments, 128–30
structure of, 55, 151
toxins in, 255
Population growth, human, 15, 33–34
Porins, 250–51
Port Aransas (Texas), 222
Portuguese man o' war, 11, 25, 79, 192
Poseidon, 7
Power plants, 9, 24, 40, 243, 245, 258
Prasher, Douglas, 116
Protista, 71
Protozoans, 11
Providence College, 87
Puerto Rico, 75–76, 78, 123
Pufferfish, 171, 190, 212, 226, 245
Puget Sound (Washington), 6, 14, 63–64, 111, 187
Purcell, Jennifer, 15–16, 33–34, 63–64, 178–79, 187, 189, 205, 299, 301–2

Qingdao (China), 15

Rabbitfish, 244–45
Red Sea, 244–45, 266, 273, 278–79, 286, 292
Red tide, 101, 303
Reefs, 5, 255, 266, 268–74, 276–81, 286, 304
artificial, 63
See also Corals
Remotely operated vehicles (ROVs), 13
Restaino, Dena, 192
Rheumatoid arthritis, 50
Rhizostoma pulmo, 51, 296. *See also* Barrel jellyfish
Rhode Island, 199
Rhopalia, 72, 122–24, 127–28, 130, 134–35, 139, 141
Rhopilema, 186, 217–20, 224, 228–30, 233, 290
Rhopilema esculentum, 43–44, 187, 217–18
Rhopilema hispidum, 44
Richardson, Tammi, 166, 173–76
Richet, Charles, 11
Rijksmuseum of Natural History (Leiden), 73–75
Rise of the Jellyfish, 57–59, 98
Robinson, Kelly, 35–36, 166, 174
Robojelly droid, 88–91
Roger Williams University, 87
Romanes, George John, xi, 131–35, 138–40, 142–45
Romans, ancient, 47
Rome, 147–48, 150, 236
Roosevelt, Theodore, 199
Rovinj (Croatia), 75
Royal Navy, British, 60

Safe Sea, 258, 263, 264, 276
Saint Helena (South Atlantic island), 199
Sakumoto, Teruko, 208, 210, 220–21, 233
Sakumoto, Yoshifumi, 209–12, 228, 233
Salento, University of, 150, 156
Salmon, 26, 90
Salps, 168–70, 303
San Andreas Fault, 137
San Diego, University of, 117
San Diego County (California), 25
San Francisco Bay area, 53–54, 95
San Pedro del Pinatar (Spain), 296
Sandwich Islands, 77
Santa Barbara (California), 24, 35, 165, 174, 195, 204
Sardine fisheries, 32, 175, 226
Saudi Arabia, 268
Science (journal), 116–17
Scotland, 24, 26–27, 60, 118, 133
Scuba, 12, 63, 98, 176, 186, 248, 276, 277

Scyphomedusae, 74
Scyphozoa, 78
Sea anemones, 5, 10–11, 65, 81, 226, 255, 259, 261–63, 274, 278
Sea cucumber, 103
Sea nettles, 7, 10, 43, 59, 61, 172–73, 192
Sea otters, 172
Sea slugs, 5, 191–92
Sea stars, 8–9, 22, 127, 170
Sea turtles, 123, 171, 194, 226
Sea urchins, 4, 5, 8, 172
Sea walnuts, 10
Sea wasp. *See Chironex fleckeri*
Sea worms, 5, 8–9, 73, 127, 168, 172, 278
Seabirds, 32, 174, 268
Seals, 32, 76
Sharks, 61, 86, 159, 168, 201, 232, 248, 252
Sherlock Holmes mysteries, 12
Sherrington, Charles, 139
Shimomura, Osamu, 110–15
Shrimp, 5, 12, 41, 121, 178, 264, 298
  acidification damage to, 8, 216–17
  flotation of, 58–59
  as food for jellyfish, 54, 121, 131, 141, 170–71
  impact of jellyfish blooms on fisheries of, 26
  jellyfishing as alternative to fishing for, 45–46, 187
  predators of, 103–5, 158
Shrine Island (Japan), 195. *See also* Miyajima
Silent Spring (Carson), 301
Siphonophores, 14, 79, 81, 104
Sisi, Abdel Fattah el-, 247
Slobodkin, Lawrence, 279
Slovenia, 77
Snails, 8, 85, 89, 161, 172, 203, 226
Sonar, 14, 45, 160
South Africa, 31, 32
South by Southwest Eco, 158
Southern California, University of, 100
Soviet Union, 27, 160
Space, experiments on jellyfish in, 128–30, 275
Space shuttle, 128, 275
Spain, 9, 26, 154, 275, 295–303
Spangenberg, Dorothy, 128–30, 275
Spearfish, 171
SpongeBob SquarePants, 44
Spotted jellyfish, 61, 64, 293
Squid, 41, 103, 106–7, 124, 159
Squirrel fish, 278
Staurozoa, 68–69, 78
Stem cells, 153–55
Stiasny, Gustav, 73–76
Stings, 24, 59, 192, 212, 300
  of corals and sea anemones, 5, 7
  of giant jellyfish, 15, 26, 33, 213, 253
  impact on fisheries, 12, 26–27, 245, 301
  incidence of, 25–26, 33, 64, 253
  lethal, 8, 26. *See also* Toxins, jellyfish
  mild, 4, 25, 44, 124, 154
  predatory function of, 72, 133, 172
  protection from, 11, 26, 243, 252, 258–59, 263–65, 291–93, 299
  by siphonophores, 14
  treatment of, 249–51, 253–54, 303
Stockholm, 118
Stolons, 151–52
Stomolophus meleagris, 44
Strobilae, *ix*, *163*. *See also* Strobilation
Strobilation, 56–57, 64, 129
Stylophora, 271
Submersibles, 13, 69, 99, 102–3, 105, 107, 168
Suez Canal, 31, 244–47, 284, 286–87
Sunfish, 171
Superstorm Sandy, 237–39
Sweden, 24
Swordfish, 171, 201
Synapses, 139–40
Syria, 30
Systema Naturæ (Linnaeus), 190

Taiwan, 46
Takahashi, Kazutoshi, 155
TED talk (Widder), 107
Tel Aviv, 4–5, 258, 276, 281, 283, 288
Tel Aviv University, 278–79, 283
Tethys (sea goddess), 56
Texas, 25, 44, 183, 210, 284–85. *See also specific cities*
Texas, University of
  at Austin, 222, 233
  at Dallas, 90
Texas A&M University at Galveston, 154
Thailand, 228
Tiberias (Israel), 258, 263, 265
Tiburonia granrojo (big red jellyfish), 99–100
Titov, Gherman, 128–29
Tokyo, 79, 225, 236
Tokyo Bay, 191

Toxins, jellyfish, 11, 234, 245, 253
  biochemistry of, 250–51, 255–56, 258
  deadliest, 26, 58, 248–50
  stinging cell mechanism for deployment of, 258–60, 263
Tripedalia, 124
Tsien, Roger, 117–19
Tsushima Island (Japan), 202, 207–14, 235, 236, 238, 297
Tuna, 159–60, 174, 225, 226
  bluefin, 171, 201
Turkey, 76, 245, 300
Turritopsis, 152–57
Twenty Thousand Leagues Under the Sea (Verne), 106, 109–10
Tyrannosaurus rex, 67

United Kingdom, 64
United Nations, 246, 286
  Food and Agricultural Organization, 147, 198, 228
United States, 10, 34, 36, 64, 100, 111, 175, 228
  atomic bombing of Japan by, 110
  carbon emissions in, 284–85
  desalination plants in, 25
  endangered animals in, 172
  jellyfish cuisine in, 48–49, 51–52
  jellyfisheries in, 228
  marine protected areas of, 199
  ports accommodating supersize ships, 31
  public aquaria in, 60
  research on medical uses of jellyfish in, 50
  in sting protocol collaborative network, 253
  television commercials in, 114
  See also specific cities, states, and regions
Uruguay, 73
USS *Ronald Reagan*, 24–25
Uye, Shin-ichi, 184, 186–89, 193, 201–2, 210, 216–25, 227–35, 299, 302

Venomous and Poisonous Marine Animals (Williamson), 44
Venus girdle, 98
Verne, Jules, 106, 109–10
Vervoort, Wim, 74
Vesuvius, 109
Vienna, University of, 73
Vietnam, 228, 299
Villanueva, Alex, 88–91
Vineyard Sound, 86, 87
Virginia Polytechnic Institute, 88
Vogel, Steven, 85

Waikiki Beach (Hawaii), 25, 250
Wales, 300
Watson, Glen, 261–63
Whales, 24, 89, 92, 106, 161, 168, 170, 229
Widder, Edith (Edie), 102–3, 105–7
Widmer, Chad, 60
Wijnhoff, Gerarda, 73, 75
Woods Hole (Massachusetts), 84–94, 96, 97, 111
  Marine Biological Laboratory (MBL), 84, 94, 111, 177
  Oceanographic Institute (WHOI), 84
World Open Water Swimming Association, 252
World War II, 73–75, 110, 160

Yamanaka, Shinya, 155, 156
Yanagawa (Japan), 216, 225–27, 229, 235
Yanagihara, Angel, 248–51, 253–55, 299
Yellow Sea, 187, 229, 234

Zappa, Frank, 156–57, 162
Zebrafish, 117
Zombie worms, 168
Zooplankton, 28, 30, 33, 76, 168, 261–62, 296